Surrogate
Humanity

PERVERSE MODERNITIES

A series edited by Jack Halberstam and Lisa Lowe

Surrogate Humanity

Race, Robots, and the Politics of Technological Futures

NEDA ATANASOSKI AND KALINDI VORA

Duke University Press   Durham and London   2019

Designed by Courtney Leigh Baker
Typeset in Whitman by Westchester Publishing Services

Library of Congress Cataloging-in-Publication Data
Names: Atanasoski, Neda, author. | Vora, Kalindi, [date] author.
Title: Surrogate humanity : race, robots, and the politics of
technological futures / Neda Atanasoski and Kalindi Vora.
Description: Durham : Duke University Press, 2019. | Series: Perverse
modernities | Includes bibliographical references and index.
Identifiers: LCCN 2018033902 (print) | LCCN 2018044223 (ebook)
ISBN 9781478004455 (ebook)
ISBN 9781478003175 (hardcover : alk. paper)
ISBN 9781478003861 (pbk. : alk. paper)
Subjects: LCSH: Robots—Social aspects. | Automation—Social aspects. |
Technological unemployment—Social aspects. | Artificial intelligence—
Social aspects. | Technological innovations—Social aspects. | Technology—
Social aspects. | Robotics—Human factors.
Classification: LCC HD6331 (ebook) | LCC HD6331 .A83 2019 (print) |
DDC 338/.0640112—dc23
LC record available at https://lccn.loc.gov/2018033902

COVER ART: Peter William Holden. "Vicious Circle," *Choreographed
Robots*. Courtesy of the artist. Photograph by Medial Mirage /
Matthias Möller.

Contents

Acknowledgments

Part of the work of collaborative scholarship is to bring forth a new subject larger than the sum of its parts. We have been friends and collaborators since we were postdoctoral scholars sharing an office in the Anthropology Department at UC Berkeley in 2007. On our daily walks to the office we discovered a shared intellectual genealogy (Neda's PhD advisor, Lisa Lowe, was a graduate of Kalindi's doctoral institution) that deeply influenced our research trajectories. We realized that our conversations about research for our first books were also generating a collaborative project on race, technology, and politics. Over the years, we have had the chance to extend our collaboration with many interlocutors who have enriched our thinking, and we are grateful to our friends, students, and colleagues for sharing news stories, provocations, and asking us difficult questions.

First and foremost, we wish to thank Lisa Lowe, who has been a friend, ally, mentor, and supporter of us and the project since its inception. Her writings on race, liberalism, and modernity, and her boundless generosity in thinking out loud with us have been impactful and inspirational, as is evident throughout the pages of this book. We are also grateful to Neferti Tadiar, who supported this project at its formative stages. Since 2012, we have had many unique opportunities to travel together and to present our work in intellectually stimulating venues. The discussions and questions that followed our talks pushed our thinking and writing in unexpected and exciting ways. Michael Dango and Rowan Bayne invited us to present our very first writing from the book at the University of Chicago Post-45

Workshop. The fruitful exchange we had with graduate students and faculty there shaped our first publication from the project that appeared in the journal *Catalyst: Feminism, Technoscience, Theory*, from which we expanded and developed a number of chapters. We are also indebted to the Technoscience Salon in Toronto, and especially its organizers Michelle Murphy and Natasha Meyers; Aimee Bahng and the Gender Research Institute at Dartmouth; Redi Koobak, Nina Lykke, and Madina Tlostanova at Linköping University in Sweden; the Digital Cultures Research Lab at Leuphana University in Lüneburg, Germany for workshopping draft chapters; Nishant Shah and Martin Warnke at Leuphana University, who invited us to present our work in Germany and allowed us the once-in-a-lifetime chance to think, write, take walks, and share meals with fellow scholars at a villa in Tuscany; Lisa Parks and her students at MIT; and Lisa Lowe and the Center for the Humanities at Tufts University.

Since 2012, many mutual friends and colleagues have provided feedback on parts of our writing and thinking. We are grateful to Banu Subramaniam, Deboleena Roy, and Angie Willey for ongoing discussions about the meaningfulness and significance of feminist collaboration. Banu continually provided us with "grist for our mill," and many of the articles she sent us along the way have made their way into the book. Lisa Cartwright and Cristina Visperas offered significant feedback and editing suggestions for our first publication from the project in *Catalyst*. We also thank Lisa Lowe, Jennifer Terry, Michelle Murphy, Julietta Hua, Xiao Liu, Don Donham, Fatima El-Tayeb, Lilly Irani, Saiba Varma, Felicity Amaya Schaeffer, Karen Barad, Lisbeth Haas, Kasturi Ray, Alex Rosenblat, Alys Eve Weinbaum, Elizabeth Losh, and Grace Hong for offering feedback or brainstorming with us about our work.

UC San Diego and UC Santa Cruz provided us with funding that enabled us to get the project off the ground in 2015, and to bring it to completion in 2018. A UCSD Academic Senate grant made possible our travel to Boston, where we conducted interviews at the MIT robotics museum and at the MIT Media Lab. We thank Kelly Dobson for an incredible tour of the space and for spending time with us during our research trip. Malathi Iyengar was our graduate student researcher who helped us to compile a rich and varied archive of Cold War–era sources on robotics, automation, and race. The UC Santa Cruz Academic Senate Special Research Grant funded Neda to travel to Germany with Kalindi and allowed us to put the finishing touches on the manuscript. At this stage, Taylor Wondergem was a graduate student researcher extraordinaire, who helped us with our citations, bibliography, images, and much, much more. She was an enthusiastic companion to us at the finish line.

Working with Duke University Press has been a pleasure. We thank Courtney Berger and Sandra Korn for their enthusiasm about the manuscript and guidance throughout the publication process. We also thank our series editors, Jack Halberstam and Lisa Lowe, for including us in their series, which we both have admired for many years. Finally, we thank the two anonymous reviewers at Duke University Press, whose suggestions for revisions made this a much better book than it otherwise would have been.

Neda

I have been fortunate to be surrounded by generous and innovative thinkers committed to a feminist and antiracist politics at UC Santa Cruz, and especially in my home department of Feminist Studies and the Program in Critical Race and Ethnic Studies. I am deeply grateful to Felicity Amaya Schaeffer, Nick Mitchell, Megan Moodie, Karen Barad, Christine Hong, and Neel Ahuja for their steadfast friendship, camaraderie, laughter, and many long conversations, and for, at various times, reading and commenting on parts of the manuscript, writing and thinking with me, or suggesting news stories and sources that made this a more thoughtful book. I am also thankful to Lisbeth Haas, Shelley Stamp, Lisa Rofel, Marcia Ochoa, Bettina Aptheker, Madhavi Murty, Irene lusztig, Adrian Brasoveanu, and Soraya Murray, whose support has sustained me as a scholar and teacher. Thinking with my graduate students has profoundly enriched my intellectual life. Erin McElroy, Yizhou Guo, Cecelia Lie, Taylor Wondergem, Francesca Romeo, Noya Kansky, Dana Ahern, Sheeva Sabati, Jess Whatcott, and Trung Nguyen have kept me inspired and excited about our work and its potential to change the world in which we live. I also thank Jennifer Suchland and the Women's, Gender and Sexuality Studies Department faculty and graduate students at the Ohio State University for their hospitality and fruitful intellectual exchanges about the project.

This book could not have been written without the many discussions, walks, hikes, drinks, meals, and writing sessions with my friends and family in Oakland and the Bay Area. I especially thank Radoslav Atanasoski, Ljiljana Atanasoska, Vesna Atanasoski, Nikki Zhang, Julietta Hua, Heidi Hoechst, Evren Savci, Morelia Portillo Rivas, Neel Ahuja, and Nick Mitchell. Elizabeth

Boschee Nagahara and her family housed me all of the many times my travels for this project took me to Boston. I am also indebted to Nathan Camp and Naruna and Kavi Vora Camp for bringing me into their family and for generously allowing me to take Kalindi away from time to time.

Kalindi

All intellectual work is collaborative, and so I thank the many colleagues and friends who have been a part of my intellectual life during the course of researching and writing this book. At UCSD, for friendship, meals, and for their important political, institutional, and scholarly work and conversation: Yen Espiritu, Jillian Hernandez, Daphne Taylor-Garcia, Lilly Irani, Fatima El-Tayeb, Saiba Varma, and Roshanak Kheshti. In the Science Studies Program, Cathy Gere, Martha Lampland, Lisa Cartwright, and Lilly Irani were fantastic colleagues in support of this project. The quarter-long "Race, Technoscience and the Speculative" working group at the Center for Humanities was an exciting forum for this research, and included Lilly Irani, Murktarat Yusef, Cristina Visperas, Curtis Marez, Roshanak Kheshti, Shelley Streeby, and Salvador Zarate. The graduate students I have worked with closely continue to inspire my thinking and commitment to working with, through, and sometimes despite the university: Linh Nguyen, Davorn Sisavath, Malathi Iyengar, Salvador Zarate, Amrita Kurian, Reema Rajbanshi, Mellissa Villa-Linton, and Hina Sheikh. New colleagues at UC Davis who have become valued colleagues in the final stages include Sara Giordano, Rana Jaleel, and Sarah McCullough. I also thank the Digital Cultures Research Lab at Leuphana University for inviting me as a senior fellow, and for their hospitality there I particularly thank Armin Beverungen, Nishant Shah, Andreas Bernard, Daniella Wentz, Paula Bialeski, and Lisa Konrad.

Life and work become entwined with political academic endeavors, and without my friends and family, none of it would be possible. For friendship, meals, hikes, playdates, and other soulful pursuits in San Diego, I thank Fatima El-Tayeb, Aleks Simic, and Blazenka Paleschek; Roshanak Kheshti, Sara Cassetti, and Diego Cassetti-Kheshti; Saiba Varma and Aftab Jassal; and Kimberly George. In this world, for their lifetime of support: Christine Vora, Alex Vora, Sonja Vora, Julie Murphy, Nathan Camp, Kavi and Naruna. In the next: Nisheeth Vora, Ann Stromberg Snell, and the other spirits who attend.

The Surrogate Human Effects of Technoliberalism

The November–December 2017 issue of *Mother Jones* playfully recasts the iconic 1932 photograph "Lunch atop a Skyscraper," which featured construction workers seated on top of a crossbeam taking a break from work to eat. The issue's cover (figure 1.1) replaces the blue-collar workers enjoying a meal together in the original image with robots wearing hard hats and overalls. Now a lone human blue-collar worker (who appears to be a white man) sits surrounded by robotic colleagues who whisper, laugh, sip coffee, and read the newspaper. The headline underneath ominously reads: "You Will Lose Your Job to a Robot—And Sooner Than You Think." Perhaps it is not coincidental that the photograph chosen for this adaptation was taken at the height of the Great Depression, when upward of 20 percent of Americans were unemployed. The article explains that within 20 years, half of current US workers will be out of jobs, and in a more distant future, even jobs that seem currently unthreatened (such as that of medical doctor) will be more efficiently—and productively—performed by robots and artificial intelligence (AI). The author speculates about this future as one that can lead to more freedom, but also more suffering, for humans:

In one sense, this all sounds great. Let the robots have the damn jobs! No more dragging yourself out of bed at 6 A.M. or spending long days on your feet. We'll be free to read or write poetry or play video games or whatever we want to do. And a century from now, this is most likely how things will turn out. Humanity will enter a golden age. But what about 20 years from now? Or 30? We won't all be out of jobs by then, but a lot of us will—and it will be no golden age. Until we figure out how to fairly distribute the fruits of robot labor, it will be an era of mass joblessness and mass poverty.[1]

This future of joblessness and poverty is echoed in the cover art of the October 23, 2017, issue of *The New Yorker* (figure I.2), which preceded the *Mother Jones* issue by just a week. A bearded young white man sits on the sidewalk, begging the passing robots for spare sprockets as they enjoy the city streets (walking their robotic dogs, looking into their smartphones, and casually sipping coffee on the go).

Though there has been a sudden increase in writing about the robot futures facing US citizens following the 2016 election of Donald Trump to the US presidency, these two covers are just a small sample of the myriad articles published in the last decade considering the advent of a robotic revolution. This is a revolution that is either celebrated as freeing humans to be less oppressed by the drudgery of wage labor, domestic and reproductive labor, the work of care, and even the work of waging war, or alternately feared as displacing humans as the masters of this world. While the inevitable incursion of robotics into domestic, social, military, and economic realms is commonly figured as a potential boon or threat to *all* of humanity, the figure of the human most threatened because it is iconically human—as the two magazine covers we've singled out vividly portray in their images (if not in the text of the articles that accompany the images)—is white and male. The human–machine future thus envisions a white loss that philosophers, politicians, and engineers must address before it is too late. Since the first industrial revolution, automation has signaled the threat of the replaceability of specific types of human functions and human workers that are racialized and gendered. This is because the tasks deemed automatable, including manual labor, blue collar factory work, and reproductive and care work, were regarded as unskilled and noncreative—work that could be done by the poor, the uneducated, the colonized, and women. Claims about the entirely novel nature of new

Figure I.1. Cover of November/December 2017 *Mother Jones* magazine.
Illustration by Roberto Parado.

Figure I.2. October 2017
New Yorker cover featur-
ing a human beggar in a
cityscape now enjoyed by
robots. Illustration by
R. Kikuo Johnson.

technologies, encapsulated by names heralding that an ostensibly unprecedented socioeconomic epoch is upon us, including the "Fourth Industrial Revolution," the "Second Machine Age," and "TechBoom 2.0," imply the ascension of humanity past differentiations in value as racialized and gendered populations.[2] This occurs due to the fact that now, even knowledge work, affective labor, and the work of taking or preserving human life have become targets for automation. Yet, even as robotics and artificial intelligence ostensibly signal universal human freedom from toil or universal human loss (of jobs, sociality, material gains, and decision-making power) as machines take over more and more tasks, questions about what kind of tasks are replaceable, and what kind of creative capacities remain vested only in some humans, indicate that humanity stands in a hierarchical if connected relationship to artificial intelligence; industrial, military, and social robots; digital technologies; and platforms that scaffold what in this book we term *technoliberalism*.

Surrogate Humanity focuses on how engineering projects that create the robots, program the AI, and enhance the digital infrastructure associated with a revolutionary new era are in fact predetermined by techniques of differential exploitation and dispossession within capitalism.[3] In this book, we propose that *technoliberalism* is the political alibi of present-day racial capitalism that posits humanity as an aspirational figuration in a relation to technological transformation, obscuring the uneven racial and gendered relations of labor, power, and social relations that underlie the contemporary conditions of capitalist production.[4] Technological futures tied to capitalist development iterate a fantasy that as machines, algorithms, and artificial intelligence take over the dull, dirty, repetitive, and even reproductive labor performed by racialized, gendered, and colonized workers in the past, the full humanity of the (already) human subject will be freed for creative capacities. Even as more valued tasks within capitalist regimes of production and accumulation, such as knowledge work, become automatable, the stated goal of technological innovation is to liberate human potential (its nonalienated essence, or core) that has always been defined in relation to degraded and devalued others—those who were never fully human. Engineering imaginaries, even as they claim revolutionary status for the techno-objects and platforms they produce to better human life, thus tend to be limited by prior racial and gendered imaginaries of what kinds of tasks separate the human from the less-than or not-quite human other.

We argue that racial logics of categorization, differentiation, incorporation, and elimination are constitutive of the very concept of technology and technological innovation. Technology thus steps into what we call a surrogate relation to human spheres of life, labor, and sociality that enables the function and differential formation and consolidation of the liberal subject—a subject whose freedom is possible only through the racial unfreedom of the surrogate. Yet there is no liberal subject outside of the surrogate–self relation through which the human, a moving target, is fixed and established. In other words, *the liberal subject is an effect of the surrogate relation*. The surrogate human effect, in this sense, is the racial "grammar" of technoliberalism. By grammar here we mean a symbolic order, following Hortense Spiller's use of the term, that establishes "feeling human" as a project of racial engineering.[5] Even as technologies like industrial, military, and companion robots are designed in ways engineers imagine will perform more perfect versions of the human—more rational killers, more efficient workers, tireless companions—such technologies still can't *feel human* in the sense that they can't feel pain or empathy. Precisely because such technologies can never be human, they allow for an exploration of the aspirations for humanity. Contrary to the seeming abandonment of the politics of difference in the so-called postrace and postlabor future projected by technoliberal discourses of machine-induced human obsolescence, we thus draw attention to the composition of the human as an abstract category whose expansive capacities continually reaffirm the racial order of things that undergirds Euro-American modernity. Put differently, the ambition to define universal humanity has been rehearsed and updated through the incorporation into engineering imaginaries of ideas about what the human is, imaginaries that guide the design of the future of the human through technologies that perform "the surrogate human effect."

In technological imaginaries both utopic (like robots that can free us from drudgery to write poetry or play video games) and paranoid (like the loss of jobs to robots), specific technologies are both actively designed, but also often feared, to act as surrogates that can free humans from having to perform historically degraded tasks. Although, in the language of science and technology studies, these technologies are coproduced with the shifting racialized and gendered essence of "the human" itself, promotional and media accounts of engineering ingenuity erase human–machine interactions such that artificial "intelligence," "smart" objects and infrastructures, and robots appear to act without any human attention. These technologies are quite explicitly termed "enchanted"—that is, within technoliberal modernity, there is a desire to attribute magic to techno-objects. In relation to the desire for enchantment, *Surrogate Humanity* foregrounds how this desire actively obscures technoliberalism's complicity in perpetuating the differential conditions of exploitation under racial capitalism.

In the desire for enchanted technologies that intuit human needs and serve human desires, labor becomes something that is intentionally obfuscated so as to create the effect of machine autonomy (as in the example of the "magic" of robot intelligence and the necessarily hidden human work behind it). Unfree and invisible labor have been the hidden source of support propping up the apparent autonomy of the liberal subject through its history, including indentured and enslaved labor as well as gendered domestic and service labor.[6] The technoliberal desire to resolutely see technology as magical rather than the product of human work relies on the liberal notion of labor as that performed by the recognizable human autonomous subject, and not those obscured labors supporting it. Therefore, the category of labor has been complicit with the technoliberal desire to hide the worker behind the curtain of enchanted technologies, advancing this innovated form of the liberal human subject and its investments in racial unfreedom through the very categories of consciousness, autonomy, and humanity, and attendant categories of the subject of rights, of labor, and of property.

Our usage of the concept of the surrogate throughout this book foregrounds the longer history of human surrogates in post-Enlightenment modernity, including the body of the enslaved standing in for the master, the vanishing of native bodies necessary for colonial expansion, as well as invisibilized labor including indenture, immigration, and outsourcing. The claim that technologies can act as surrogates recapitulates histories of disappearance, erasure, and elimination necessary to maintain the liberal subject as the agent of historical progress. Thus, framing the surrogate human effect as the racial grammar of technoliberalism brings a feminist and critical race perspective to bear on notions of technological development, especially in the design and imagination of techno-objects and platforms that claim to reenchant those tasks understood as tedious or miserable through the marvels of technological progress—ostensibly dull, dirty, repetitive, and uncreative work.

To understand how claims of human freedom and human loss enabled by technological development allow for the retrenchment of the liberal subject as the universal human, *Surrogate Humanity* foregrounds the obfuscated connections between the human–machine divide in US technological modernity and the racial production of the fully human in US political modernity. Focusing on the material, social, and political consequences of the mutual generation of "the human" and "the machine" from the US post–World War II standardization of automation into the present, we explore both the social impact of design and engineering practices intended to replace human bodies and functions with machines *and* the shift in the definition of productivity, efficiency, value, and "the racial" that these technologies demand in their relation to the post-Enlightenment figure of the human. We begin with the second half of the twentieth century because this is the moment when the United States ascends to global political and economic supremacy and cultural influence, inheriting the mantle of its own and Western European settler imperial social structures. At this same historical juncture, the racial architecture of US modes of governance and geopolitical ascendancy were erased in the logics of post–civil rights racial liberalism and multiculturalism.[7] Crucially, the advent of what can be termed, ironically, a "postracial" domination translates directly into the perception of new technologies as neutral and disembodied, even as these technologies are anchored in, and anchor, contemporary US imperial power. In short, the technological sphere has been separated from the racial scaffolding of the social in the Cold War and post–Cold War eras. Yet, as we argue, it is essential to assess the racial and gendered architecture of post-Enlightenment modernity as engineered into the form and function of given technologies. This calls for situating techno-objects and platforms in a social relation to what is experienced as a "human." Thus, although our book is primarily focused on present-day claims about the revolutionary nature of new digital technologies, robotics, and AI, throughout our analysis of techno-objects and the social and political discourses that frame them, we unearth the obscured histories that delimit technoliberal engineering projects focused on efficiency, productivity, and further accumulation through dispossession.

Throughout this book, we insist on the infusion of a seemingly neutral technological modernity with the racial, gendered, and sexual politics of political modernity, based as they are in racial slavery, colonial conquest and genocide, and forced mobility through ongoing racial imperial practices of labor allocation and warcraft. To accomplish this, we extend critical ethnic studies analyses of gendered racialization to include machine "others." By focusing on machines, we take the weight of an ethnic studies analysis off of racialized people so that we can see how this relationship functions even in their absence. Tracking the surrogate human effect within technoliberal politics enables us to attend to techniques through which difference (whether human–nonhuman or interhuman) is produced, while understanding categories of difference as historically specific.

By tracking how the surrogate human effect functions as the racial grammar of technoliberalism, we connect critiques of historical and political consciousness, freedom, and agency, whether of the machine or of the liberal subject, to calls for thinking beyond the limits of liberal humanist visions of more just futures. We thus position our critique of technoliberalism in relation to how technologies can be used to create relations between the human and the machine that are outside of the use–value–efficiency triad of capitalist modes of production. We see this work of redescribing value, and what or who is valuable, outside of the parameters of racial capitalism and its modes of waging war and staging social relations already happening in artistic and engineering projects focused on creating technologies that blur the boundaries between subject and object, the productive and unproductive, and value and valuelessness, thereby advancing structures of relation that are unimaginable in the present. Pushing against the limits of the imagination imposed by the symbolic logics of the surrogate human effect, the artistic, literary, engineering, and scientific projects we include in juxtaposition with those we critique refuse existing frames for recognizing full humanity, particularly the categories of the liberal politics of recognition such as the subject of labor or human rights.

The Surrogate Human Effect

Like the "others" of the (white) liberal subject analyzed by decolonial and postcolonial scholarship, the surrogate human effect of technology functions first to consolidate something as "the human," and second to colonize "the human" by advancing the post-Enlightenment liberal subject of modernity as universal.[8] The concept of the surrogate brings together technoliberal claims that technological objects and platforms are increasingly standing in for what the human does, thus rendering the human obsolete, while also foregrounding the history of racial unfreedom that is overwritten by claims of a postrace and postgender future generated by that obsolescence. In our usage, the longer history of the surrogate human effect in post-Enlightenment modernity stretches from the disappearance of native bodies necessary for the production of the fully human, through the production of the fungibility of the slave's body as standing in for the master, and therefore also into the structures of racial oppression that continue into the postslavery and post–Jim Crow periods, and into the disavowal of gendered and racialized labor supporting outsourcing, crowdsourcing, and sharing economy platforms. Framing technologies through the lens of the surrogate effect brings a feminist and critical race perspective to bear on notions of technological development, especially in the design and imagination of techno-objects and platforms that claim a stand-in role for undesirable human tasks.

As part of the surrogate effect, the surrogate is a racialized and gendered form defining the limits of human consciousness and autonomy. Saidya Hartman conceptualizes the surrogate by citing Toni Morrison's formulation of slaves as "*surrogate selves* for the meditation on the problems of human freedom."[9] Hartman proposes that "the value of blackness resided in its metaphorical aptitude, whether literally understood as the fungibility of the commodity or understood as the imaginative surface upon which the master and the nation came to understand themselves."[10] The slave, the racialized fungible body, also acts as a "surrogate for the master's body since it guarantees his disembodied universality and acts as the sign of his power and domination."[11] As Hartman elaborates, these racialized structures of the surrogate did not simply disappear after emancipation. Rather, "the absolute dominion of the master, predicated on the annexation of the captive body, yielded to an economy of bodies, yoked and harnessed, through the exercise of autonomy, self-interest, and consent. . . . Although no longer the extension and instrument of the master's absolute right or dominion, the laboring black body remained a medium of others' power and representation."[12]

While Hartman is referencing the rise of new modes of bonded labor following emancipation that were encapsulated by the liberal formalities of contract, consent, and rights, her theorization of surrogacy as a racialized and gendered arrangement producing autonomy and universality of and for the master is useful for thinking about the contemporary desire for technology to perform the surrogate human effect. The racialized and gendered scaffolding of the surrogate effect continues to assert a "disembodied universality" that actually offers the position of "human" to limited human actors, thereby guaranteeing power and domination through defining the limits of work, violence, use, and even who or what can be visible labor and laboring subjects.

Tracking the endurance of the racial form of slavery as the (not so) repressed or spectral frame for the imaginary of what surrogate technologies do, or who or what they are meant to replace, we insist throughout this book that human emancipation (from work, violence, and oppressive social relations) is a *racialized aspiration for proper humanity* in the post-Enlightenment era. In the US context, reading technologies as they reflect the dominant imagination of what it means to be a human thus means that they are situated in social relations of race, gender, and sexuality, as these derive from embodied histories of labor, Atlantic chattel slavery, settler colonialism, and European and US imperialism, to name the most dominant. The preeminent questions of the politics of the subject, and the derivative politics of difference that consume critical theory—questions that are about political consciousness, autonomy with its attendant concepts of freedom and unfreedom, and the problem of recognition—also drive the preeminent questions we must ask of technologies that perform the surrogate human effect.

The surrogate effect of technological objects inherits the simultaneously seeming irrelevance yet all-encompassing centrality of race and histories of enslavement and indenture against which the liberal subject is defined. As Lisa Lowe writes:

> During the seventeenth to nineteenth centuries, liberal colonial discourses improvised racial terms for the non-European peoples whom settlers, traders, and colonial personnel encountered. We can link the emergence of liberties defined in the abstract terms of citizenship, rights, wage labor, free trade, and sovereignty with the attribution of racial difference to those subjects, regions, and populations that liberal doctrine describes as unfit for liberty or incapable of civilization, placed at the margins of liberal humanity.[13]

Lowe explains that while it is tempting to read the history of emancipation from slave labor as a progress narrative of liberal development toward individual rights and universal citizenship, in fact, "to the contrary, this linear conception of historical progress—in which the slavery of the past would be overcome and replaced by modern freedom—concealed the persistence of enslavement and dispossession for the enslaved and indentured" and racialized populations necessary to the new British-led imperial forms of trade and governance "expanding across Asia, Africa, and the Americas under the liberal rubric of free trade."[14] Moreover, according to Lowe, "the liberal experiment that began with abolition and emancipation continued with the development of free wage labor as a utilitarian discipline for freed slaves and contract laborers in the colonies, as well as the English workforce at home, and then the expanded British Empire through opening free trade and the development of liberal government."[15] While the history of capitalism tends to be written as the overcoming of serf, slave, and indentured labor through free contract and wage labor, that is, as freedom overcoming unfreedom, as Lowe demonstrates, it is actually the racialized coupling of freedom and unfreedom that undergird and justify capitalist and imperial expansionism.

Rather than freedom being on the side of modernity, which overcomes the unfreedom that is the condition of premodernity, in fact the states of both freedom and unfreedom are part of the violent processes of extraction and expropriation marking progress toward universality. Undergirding Euro-American coloniality, political liberalism maintains the racial temporality of post-Enlightenment modernity that depends on innovating both bodies and resources (and how each will be deployed). David Theo Goldberg argues that liberalism is the "defining doctrine of self and society for modernity," through which articulations of historical progress, universality, and freedom are articulated.[16] Because liberalism's developmental account of Euro-American moral progress has historically been premised on the transcending of racial difference, as Goldberg puts it, under the tenets of liberalism, "race is irrelevant, but all is race."[17]

To articulate freedom and abstract universal equality as the twin pillars of liberal modes of governance, racial identity categories and how they are utilized for economic development under racial capitalism are continually disavowed even as they are innovated. In her writing about how such innovations played out in the post–World War II context, the historical period in which we locate our study, Jodi Melamed has argued that US advancement toward equality, as evidenced by liberal antiracism such as civil rights law and the professional accomplishments of black and other minority citizens, was meant to establish the moral authority of US democracy as superior to socialist and communist nations.[18] Highlighting antiracism as the central tenet of US democracy, the US thus morally underwrote its imperial projects as a struggle for achieving states of freedom abroad over illiberal states of unfreedom, racializing illiberal systems of belief as a supplement to the racialization of bodies under Western European impe-rialism.[19] The assertion that the US is a space of racial freedom, of course, covered over ongoing material inequalities based on race at home. As part of the articulation of US empire as an exceptional empire whose violence is justified because it spreads freedom, the history of slavery is always ac-knowledged, but only insofar as it can be rendered irrelevant to the present day—that is, the history of slavery is framed as a story of US national over-coming of a past aberrant from the ideals of US democracy, and as a story of redemption and progress toward an inclusion as rights-bearing subjects of an ever-proliferating list of others (women, black people, gay people, disabled people, etc.).

Technoliberalism and Racial Engineering of a "Post"-Racial World

"Will robots need rights?" This dilemma was included in *Time* magazine's September 21, 2015, issue as one of the most important questions facing US society in the present day. In his written response, Ray Kurzweil, an inven-tor and computer scientist, wrote that "If an AI can convince us that it is at human levels in its responses, and if we are convinced that it is experiencing the subjective states that it claims, then we will accept that it is capable of experiencing suffering and joy," and we will be compelled to grant it rights when it demands rights of us.[20] In other words, if a robot can prove that it can feel human (feel pain, happiness, fear, etc.), its human status can be recognized through the granting of rights. Philosophical and cultural medi-tations upon questions of artificial personhood, machinic consciousness, and robot autonomy such as that in *Time* magazine announce the advent of what we term in this book *technoliberalism* by asserting that in the current moment, the category of humanity can be even further expanded to poten-tially include artificial persons. According to Hartman, under liberalism, the "metamorphosis of 'chattel into man'" occurs through the production of the liberal individual as a rights-bearing subject.[21] However, as Hartman elaborates, "the nascent individualism of the freed designates a *precarious autonomy since exploitation, domination, and subjection inhabit the vehicle of rights*."[22]

Autonomy and consciousness, even when projected onto techno-objects that populate accounts of capitalist futurity, continue to depend on a racial relational structure of object and subject. We describe this symbolic ordering of the racial grammar of the liberal subject the "surrogate human effect." As technology displaces the human chattel-turned-man with man-made objects that hold the potential to become conscious (and therefore autonomous, rights-bearing liberal subjects freed from their exploitative conditions), the racial and gendered form of the human as an unstable category is further obscured. Technoliberalism's version of universal humanity heralds a postrace and postgender world enabled by technology, even as that technology holds the place of a racial order of things in which humanity can be affirmed only through degraded categories created for use, exploitation, dispossession, and capitalist accumulation. As Lisa Lowe articulates, "racial capitalism suggests that capitalism expands not through rendering all labor, resources, and markets across the world identical, but by precisely seizing upon colonial divisions, identifying particular regions for production and others for neglect, certain populations for exploitation, and others for disposal."[23] As we show throughout the chapters of this book—which range in scope from examining how technological progress is deployed as a critique of white supremacy since the advent of Trumpism, effectively masking how the fourth industrial revolution and the second machine age have accelerated racialized and gendered differentiation, to how the language of the sharing economy has appropriated socialist conceptions of collaboration and sharing to further the development of capitalist exploitation—within present-day fantasies of techno-futurity there is a reification of imperial and racial divisions within capitalism. This is the case even though such divisions are claimed to be overcome through technology.

Surrogate Humanity contends that the engineering imaginaries of our technological future rehearse (even as they refigure) liberalism's production of the fully human at the racial interstices of states of freedom and unfreedom. We use the term *technoliberalism* to encompass the techniques through which liberal modernity's simultaneous and contradictory obsession with race and its irrelevance has once again been innovated at the start of the twenty-first century, with its promises of a more just future enabled by technology that will ostensibly result in a postrace, postlabor world. This is also a world in which warfare and social relations are performed by machines that can take on humanity's burdens. Technological objects that are shorthand for what the future should look like inherit liberalism's version of an aspirational humanity such that technology now mediates the freedom–unfreedom dynamic that has structured liberal futurity since the post-Enlightenment era. Put otherwise, technoliberalism proposes that we are entering a completely new phase of human emancipation (in which the human is freed from the embodied constraints of race, gender, and even labor) enabled through technological development. However, as we insist, the racial and imperial governing logics of liberalism continue to be at the core of technoliberal modes of figuring human freedom. As Ruha Benjamin puts it, "technology . . . is . . . a metaphor for innovating inequity."[24] To make this argument, she builds on David Theo Goldberg's assessment of postraciality in the present, which exists "today alongside the conventionally or historically racial. . . . In this, it is one with contemporary political economy's utterly avaricious and limitless appetites for the new."[25] Yet amid assertions of technological newness, as Benjamin demonstrates, white supremacy is the default setting.

Technoliberalism embraces the "post"-racial logic of racial liberalism and its conception of historical, economic, and social newness, limiting the engineering, cultural, and political imaginaries of what a more just and equal future looks like within technological modernity. As we propose, race and its disciplining and governing logics are engineered into the form and function of the technological objects that occupy the political, cultural, and social armature of technoliberalism. Rather than questioning the epistemological and ontological underpinnings of the human, fantasies about what media outlets commonly refer to as the revolutionary nature of technological developments carry forward and reuniversalize the historical specificity of the category *human* whose bounds they claim to surpass.

Our book addresses not just how technologies produce racialized populations demarcated for certain kinds of work, but also how race produces technology in the sense that it is built into the imaginaries of innovation in engineering practice.[26] To do so we build on and expand on the work of scholars like Wendy Chun and Beth Coleman, who have proposed thinking about race as technology. Chun demonstrates that conceptualizing race as a technology (not as an essence, but as a function) lets us see how "nature" and "culture" are bound together for purposes of differentiating both living beings and things, and for differentiating subjects from objects.[27] This formulation allows us to trace the conceptual origins of race as a political category rooted in slavery and colonialism that has enduring legacies (both in terms of classifying people and in terms of inequities). Similarly, Beth Coleman argues that conceptualizing race as a technology highlights the productive work that race does (as a tool, race can in some contexts even be seen to work in ways that are separable from bodies).[28] While such reconceptualizations of race as a category are valuable, they do not fully account for race as the condition of possibility for the emergence of technology as an epistemological, political, and economic category within Euro-American modernity. As such, technology undergirds the production of the human as separate from the machine, tool, or object. Technology is a racial category in that it reiterates use, value, and productivity as mechanisms of hierarchical differentiation and exploitation within racial capitalism.

Our focus on race and gender, and freedom and unfreedom, within the technoliberal logics that configure the aspirational temporality of feeling human in the twenty-first century brings a critical race and ethnic studies perspective to the imaginary of historical progress that pins hopes for achieving universal human freedom on technological development. Decolonial thought, critical race studies, and feminist science studies, each of which has differently engaged post- and antihumanism to extend an analysis of the vitality and agency of objects and matter to problematize the centrality of modern man in the field of the political, can thus productively be put into dialogue as a starting point for theorizing technology beginning with difference. According to Alexander Weheliye, "the greatest contribution to critical thinking of black studies—and critical ethnic studies more generally . . . [is] the transformation of the human into a heuristic model and not an ontological fait accompli."[29] Weheliye argues that, given developments in biotechnology and informational media, it is crucial to bring this critical thought to bear upon contemporary reflections on the human.[30] As is well known, eighteenth- and nineteenth-century European colonialism, a structure that instituted a global sliding scale of humanity through scientific notions about racial differences and hierarchies, undergirded systematic enslavement and subjugation of nonwhite peoples to advance European capitalism and the industrial revolution. Developed alongside and through the demands of colonialism, this scale designated a distinction among human beings, not just between humans and animals, such that humanity was something to be achieved.[31] Decolonization, Frantz Fanon wrote, is in this respect "quite simply the replacing of a certain 'species' of men by another 'species' of men."[32] At stake in the Fanonian concept of decolonial revolution is the reimagining of the human–thing relation as a precondition for freedom. This is precisely the relation that the techno-revolutionary imaginary scaffolding technoliberalism fails to reenvision. This failure is due in part to the fact that, at the same time that colonialism was without a doubt a project of dehumanization, as scholars like David Scott and Samera Esmeir show, European colonialism through its discourses of technological innovation, progress, and civilization also aimed to "humanize" racialized others.[33]

Engineering imaginaries about technological newness that propose to reimagine human form and function through technological surrogates taking on dull, dirty, repetitive, and reproductive work associated with racialized, gendered, enslaved, indentured, and colonized labor populations thus inherit the tension between humanization and dehumanization at the heart of Western European and US imperial projects. On the one hand, there is a fear that as technologies become more proximate to humans, inserting themselves into spheres of human activity, the essence of humanity is lost. On the other hand, the fantasy is that as machines take on the sort of work that degrades humans, humans can be freer than ever to pursue their maximum potential. As we postulate, this tension arises because even as technoliberalism claims to surpass human raced and gendered differentiation, the figuration of "humanity" following the post- of postracial and postgender brings forward a historically universalizing category that writes over an ongoing differential achievement of the status of "the human."

In contrast to speculative writing that recent developments in robotics and AI can liberate humanity by ending the need for humans to perform degraded, dull, dirty, or repetitive tasks, decolonial and critical race scholars such as Sylvia Wynter first ask who or what falls into and out of the category of human, signaling that the human as a shifting and disciplining category continues to be profoundly racialized, and only then poses the question of what sorts of redescriptions of the human are necessary to conceive of what comes "after Man." To paraphrase Wynter, in order to wrest a new figure of the human (or, less ambitiously, a new human potentiality) from the epistemological break that follows from the techno-revolution, we must unmake the world in its current descriptive-material guise.[34] Wynter's call for redescribing the human after-Man as an ontological problem points to the coexistence of the world of the humanist subject (Man) with those other worlds forcibly written upon by colonial practices that continue outside/alongside it.[35] To get at the problem of how the category of the human is constituted through material histories of difference, Donna Haraway rejects "human exceptionalism" and instead centers the "encounter" between humans and nonhumans and between subjects and objects, as constitutive of who and what is encountered.[36] Similarly, when the material world is viewed through Karen Barad's analytic of intra-activity,[37] the centrality of the liberal human subject can be suspended. The historically conditioned (racialized and gendered) nature of subject–object and human–thing encounters, we argue, is precisely what technoliberal imaginaries overwrite through an emphasis on the seeming neutrality and rationality of technoliberal futurism.

The Enchanted Future of Technoliberal Modernity and the Racial Conditions of Freedom

As we have proposed thus far, a core aspect of the surrogate effect as the racial grammar of technoliberalism is the articulation of progress tethered to present-day technologies, including their "intelligence," internetworking, and engineering. Terms that mark our ostensibly new technological epoch, such as fourth industrial revolution and the second machine age, posit our economic arrangements and social infrastructures as nothing short of revolutionary—that is, entirely different from those that led to the first industrial revolution and earlier moments of automation. Smart internetworked objects like the Roomba self-directing vacuum cleaner, prescription pill bottles that automatically order refills when their volume gets low, or umbrellas that light up with the prediction of rain, seem to intuit human needs. They are imagined to promise a future in which animate objects manage the dull work of reproducing daily existence.[38] MIT Media Lab instructor and entrepreneur David Rose has termed such technologies "enchanted objects."

The desire for technological enchantment, that is, for animate and "intelligent" technological objects that perform degraded and devalued tasks to uphold the freedom of the liberal subject, perpetuates the surrogate effect of technoliberalism, erasing the ongoing ways in which the colonial structures of racialized and gendered exploitation that enable the feeling of being human produce the desire for enchanting technology. Throughout the chapters of Surrogate Humanity, we dwell on the tension between economic and technological rationality, the hallmarks of political and economic liberalism, and the engineering and cultural imaginaries that seek to reenchant our technological modernity through machines,

Figure I.3. Enchanted broom from Disney's 1940 film *Fantasia*.

platforms, and apps whose magic is to remove human exploitation from the frame. Put otherwise, the technoliberal fantasy of a reenchanted secular and rational world is one made magic through technologies that can be completely controlled by humans, yet these same technologies bypass human thought and labor, thereby seeming to overcome the historical, economic, and imperial legacies that create categories of objects and people as needed, desired, valuable, or disposable. Enchanting the object precludes the possibility of recognizing the racialized and gendered scaffolding of racial capitalism and of an attendant antiracist politics. A desire for enchanted objects extends from European-derived fairy tales and Disney's animated films such as *Fantasia* ("The Sorcerer's Apprentice," figure 1.3) and *Cinderella*, in which ensorcelled wash pails, dust brooms, and pumpkins free the apprentice and the orphan from their toils, to contemporary Euro-American fictional texts, including *Harry Potter* and *Lord of the Rings*, that feature extraordinary objects like swords that anticipate the enemy.[39] These fantasies are about emancipation from manual, repetitive, and unimaginative labor by turning the tools of work into the worker as pails and brooms (or the modern-day Roomba) move on their own. They thus extend the history of the autonomous subject whose freedom is

Figure I.4. "Enchanted Objects: Organizing the Internet of Things by Human Desires."
Poster by Jean-Christophe Bonis.

in actuality possible only because of the surrogate effect of servants, slaves, wives, and, later, industrial service workers who perform this racialized and gendered labor (see figure 1.4).

Technological enchantment seeks to overcome a sense of disappointment in the limitations of the human as a biological being embedded in a rational-secular-scientific society. In this future imaginary, human consciousness shifts vis-à-vis the technical enchantment of objects, animate and artificially intelligent, rather than as a result of political transformations.

The "smartness" of smart objects brings artificial intelligence to the center of the enchantment of technology. The question "What is intelligence?" undergirds the desire for the enchantment of technological modernity via humanlike, but not quite human, objects, and is informed by the history of debates and developments in artificial intelligence since the middle of the twentieth century. In the seminal 1961 article "Steps toward Artificial Intelligence," Marvin Minsky, the cognitive scientist and cofounder of MIT's Artificial Intelligence Laboratory, argued that "intelligence seems to denote little more than the complex of performances which we happen to respect, but do not understand."[40] For Minsky, the fact that human beings do not understand intelligence did not mean that machines could not think. Rather, Minsky argued that when machines are equipped with inductive reasoning and a model of the universe, "or an ensemble of universes, and a criterion of success," the problem of intelligence becomes technical rather than philosophical.[41]

In Minsky's model universe in which artificially intelligent creatures act, the creatures' self-representation depends upon and reiterates a Cartesian mind–body split. Minsky explains that "our own self-models have a substantially 'dual' character; there is a part concerned with the physical or mechanical environment . . . and there is a part concerned with social and psychological matters. It is precisely because we have not yet developed a satisfactory mechanical theory of mental activity that we have to keep these areas apart."[42] Given these models, even a robot, when asked "what sort of being it is," must respond "by saying that it seems to be a dual thing—which appears to have two parts—a mind, and a body."[43] In this sense, Minsky's proposition is in line with what N. Katherine Hayles has argued in relation to the post–World War II history of informatics and cybernetics, namely that

> the erasure of embodiment is a feature common to *both* the liberal humanist subject and the cybernetic posthuman. Identified with the rational mind, the liberal subject *possessed* a body but was not usually represented as *being* a body. Only because the body is not identified with the self is it possible to claim for the liberal subject its notorious universality, a claim that depends on erasing markers of bodily difference, including sex, race, and ethnicity. . . . To the extent that the posthuman constructs embodiment as instantiation of thought/information, it continues the liberal tradition rather than disrupts it.[44]

In contrast to the work in artificial intelligence by Minsky, the roboticist and former director of the MIT Artificial Intelligence laboratory, Rodney Brooks, made the body of the robot central to intelligence. In Brooks's version, the enchantment of the technological object is made manifest because the robot's physical presence in the world allows it to learn without human programming. In other words, the "magic" of the robot is that it can learn without human intervention. In Brooks's own words, "for a machine to be intelligent, it must draw on its body in that intelligence."[45] This version of technological smartness makes the physical form of the robot primary. To accomplish this, Brooks's engineering took a radically different turn from traditional robotics research in the 1980s when he took the cognition box out of his robots.[46] The cognition or computational box, what had been thought of as "the heart of thinking and intelligence" in a robot, served the purpose of instructing the machine in "what computations it should do, and how much feedback should go into the perceptual process and come from the motor process."[47] As Brooks writes in *Flesh and Machines*, leaving out the cognition box so that the machine would exclusively focus on sensing and action left out what was traditionally thought of as "intelligence" in AI.[48] He explains that it is not that he was "giving up on chess, calculus, and problem solving as part of intelligence" by leaving out the cognition box; rather, his "belief at the time, and still today, is that [these sorts of intelligence actually] arise from the interaction of perception and action."[49]

Whereas computer intelligence without a body marks one mode of enchanting technology that removes the human, in Brooks's version of what defines a robot, the human is once again removed, here in the sense that the programmer is removed and the robot (as if enchanted) can learn on its own through moving around the world and perception. Brooks's students, including Cynthia Breazeal, whose work we discuss in detail in chapter 4, go so far as to explicitly remark that robots are magic.[50] Science and technology studies (STS) scholar Lucy Suchman persuasively establishes that in the history of sociable robotics emblematized by Brooks and Breazeal, ideas about "the world" and the bodies that move around in that world are given, and that, therefore, "the world" is seen as a fixed and pre-given entity.[51] Removing the human from public accounts of how robots move about and act in the world reaffirms autonomy and rational agency as the two key attributes of personhood, which are culturally specific to post-Enlightenment Europe.[52] Suchman reminds us that it is only when human labor and its alignments with nonhuman components are made invisible that a seemingly autonomous technology can come to exist.[53]

The desire for enchanted "smart" technologies (both embodied and disembodied) points to the desire for objects to perform the surrogate effect that reaffirms post-Enlightenment conceptions of human autonomy, and therefore freedom, as separate from "things" that are intended for pure use. It is in this mode that the enchanted object allows the liberal subject to feel human. Hortense Spillers details the history of how the liberal subject of US racial capitalism realizes its freedom only through its desire for the "irresistible, destructive sensuality" of the captive body.[54] Spillers explains that the captive body is defined by externally determined "meaning and uses." Sex robotics engineering, discussed in the epilogue, provides an illustrative example of this desire. "Silicone Samantha," a prototype sex robot being developed by RealBotix, can be controlled by users shifting between more and less sexually responsive modes. RealBotix plans to "enhance user plea-sure" through the robotic simulation of reciprocity, desire, and pleasure (orgasm). The desire for an enchanted object of sexual satisfaction reminds us of the historical imprint of that desire tracked in Spillers's analysis of the mark of racial slavery upon the liberal subject. The freedom of this subject conditions it to desire a subjectless "other" who "embodies sheer physical powerlessness."[55] The imprint is a desire that knows its own freedom only through the complete domination of the object of its pleasure, even when, and perhaps especially when, that body can simulate what is in fact an inscrutable, because dominated, pleasure or reciprocity.

The technoliberal desire driving the development of sex robotics moves the ordering of things that occurs through historical structures of racialized desire into the realm of engineering imaginaries, enacting the surrogate ef-fect of technology as the racial grammar of technological enchantment. Michel Foucault's *History of Sexuality* and *The Order of Things* elaborate how the ordering of modernity occurs through the reification of categories through which subjects and knowledge are ordered, with desire being one of the principles organizing the subject. Ann Laura Stoler has shown how colonialism folded racialized structures of desire and power between colo-nizers and the colonized into the gendered domestic sphere.[56] Technolib-eral desire extends these structures into the sphere of a growing technical infrastructure of robotics, algorithms, platforms, and interfaces examined in this book.

Contrary to bringing about a seemingly enchanted world in which ab-stract equality and the end of human exploitation have been achieved as the result of technological development, new technologies that automate not just industry, but desire and emotion, further shroud the racial and gendered dynamics that have historically obscured the physical and affec-tive work involved in supporting the production of the fully human subject. Moreover, technological modernity has historically ramped up production and the need for more workers whose labor has been devalued due to auto-mation, as, for instance, in twentieth-century agribusiness.[57] We thus agree with feminist and critical race scholars of STS like Banu Subramaniam and Rebecca Herzig, among others, who insist that what is needed today is an exploration of the ways in which geopolitical and economic shifts demand a grappling with new subjects of labor, including, for instance, nonagential labor, animal and nonhuman labor, and metaphysical labor, which remain unrecognized as laboring subjects in current scholarly and public discussions about the future of work.[58] At the same time, we point throughout the book to the limits of the category of labor as a politics of recognition and visibility to fully disrupt technoliberal logics, as we also do with the subject categories produced by rights and property.

Dominant techno-utopic imaginaries direct funds and structure engineering research labs around the world, and therefore also impact the distribution of differential conditions of comfort versus misery in the present along vectors of race, gender, class, and other social hierarchies. The surrogate human effect explains how difference continues to inform what subjects become legible as human through technology design imaginaries that respond to market values by focusing on innovating and improving, rather than challenging, social and cultural structures and processes that are predicated by categories of gendered racial hierarchy. To this end, Denise da Silva offers the concept of "knowing (at) the limits of justice," a practice that "unsettles what has become but offers no guidance for what has yet to become."[59] To insist on "knowing at the limits" of representational categories of difference, we must ask: If the predominant fantasies of systemic social change in mainstream Euro-American public discourse dwell upon the techno-utopics of a world in which all of those who are already human and already subjects ascend into the realm of those whose lives are supported by "human-free" or "unmanned" technological infrastructures of service (whether in factories, in the military, or in the nursing home), then how do we think about the relationship of new technologies to possible fields of political protest or action?

The dissident technological imaginaries we include in each chapter take up categories that challenge those of technoliberal capitalism and its projected futures. We read these design imaginaries as exploring the possibilities of technology to break from historically sedimented dynamics of freedom and unfreedom woven into the fabric of technological modernity. In addition to offering critique, each chapter thinks through how such design imaginaries can push at the limits of what is possible, disrupting the confining notions of (technoliberal capitalist) possibility housed in the engineering imaginaries we critique. We explore these questions through juxtaposing engineering imaginaries that embrace the surrogate effect, thereby advancing the infrastructure of technoliberal futures, with imaginaries that do not.

Using examples of robotic technologies intended to replace human bodies and functions from the early twentieth century to the present day, the first chapter foregrounds the postlabor and postrace imaginary of present-day "technoliberalism" as a reinvigoration of the historical imbrications of liberalism and fascism—the twin pillars of US economic, social, and geopolitical supremacy. Rather than posit a break between the liberal and fascist logics of automation, we insist on their codependence. We survey the ways in which automation in both the liberal-capitalist and totalitarian-fascist bents depends upon a fantasy of robotics tied to the history of racial slavery and the myth of a worker who cannot rebel. We track this foundational fantasy through Cold War discourses of automation as mediating the distinction between democratic liberalism and totalitarianism as the prehistory of contemporary discourses around robotics and white loss in the era of the Trump presidency.

Building on our analysis of how liberalism and fascism have deployed and constructed fantasies of the fully human through and against capitalist logics of automation, the second chapter turns to present-day technoliberalism's appropriation of socialist imaginaries of the commons, sharing, and collaboration. These three terms have become the buzzwords used to describe the economic transformations marking the so-called fourth industrial revolution and second machine age. While making claims to radical shifts toward an economy where commodities can be shared, and where 3D printers can even lead to the end of capitalism as we know it, as we argue, such technoliberal progress narratives in fact mask the acceleration of exploitation under the conditions of racial capitalism. Critiquing such appropriative moves in collaborative robotics, the sharing economy, and the creative commons, we also read alternate genealogies and visions of collaboration, sharing, and technology in collectivist and decolonial feminisms.

In the next chapter, we extend this discussion of the acceleration of exploitation by turning our attention to the ways in which claims that technology is displacing human labor invisibilize the growing workforce of casualized and devalued laborers performing tasks that we are encouraged to imagine as performed for us by robots and AI. Addressing the relationship between service and the promises of technoliberal futurity, we assess how present-day disappearances of human bodies take place through platforms specifically designed to disguise human labor as machine labor. Focusing on the labor politics, design, and infrastructures of service, we argue that platforms like Alfred and Amazon Mechanical Turk enact the surrogate effect for consumers through the erasure of human workers. Consumers therefore consume the assurance of their own humanity along with the services provided.

Following from this discussion of the erasure of the potential physical and social encounter between worker and consumer through digital platforms, chapter 4 turns to robots that are designed to take up a different kind of social relation with the human: so-called sociable emotional robots. We argue that machine sociality preserves the effect of human uniqueness, as the social function of the robot is continually reduced to service performed through the correct display of obeyance and eager responsiveness to human needs. Focusing on the robot Kismet, which is considered the first sociable emotional robot, we draw attention to the imperial and racial legacies of a Darwinian emotion-evolution map, which was the model for Kismet's emotional drives. We analyze how sociable emotional robots are designed as a mirror to prove to us that the apex of human evolution, resulting from these racial legacies, is the ability to perform the existence of an interior psyche to the social world.

The next two chapters continue the discussion of service, human–machine relations, and the technoliberal racial engineering of robotics in the automation of warfare. Chapter 6 addresses drones (semiautonomous weapons) and so-called killer robots (autonomous lethal weapons) as technologies that conjure the dangerous specter of machine autonomy in US public debates about the potential threat to humanity posed by AI. This chapter contends with the configuration of autonomy within military technologies that produces killable populations as "targets," and builds on post-Enlightenment imperial tools of spatial and temporal command to refigure contemporary warfare as "unmanned." We assert that both autonomous and semiautonomous weapons are in fact not "unmanned," but cobots, in the sense that they are about human–machine coproduction. The chapter thus problematizes conceptions of autonomy that at once produce myths of unmanned warfare and the racialized objecthood tethered to servitude within technoliberalism.

The final chapter elaborates our analysis of how speculation about the future of lethal autonomous weapons engenders present-day fears around machine autonomy in ways that continue to conceive historical agency in relation to the racialized inheritances defining objecthood, property, and self-possession. We argue that the killer robot is a technology that enables a description of what it means to feel human within technoliberal imperialism. To do so, we turn to attempts by human rights organizations and NGOs to ban killer robots (autonomous weapons that could make decisions about taking human life without human oversight). These groups argue that killer robots are a human rights violation in the future tense, since fully autonomous lethal weapons are not currently operational in the field of war. Against the specter of the killer robot as an a priori human rights violation, humanity is rendered as the capacity to feel empathy and recognize the right to life of killable others, while reifying the human as the rights-based liberal subject.

1. Technoliberalism and Automation

Racial Imaginaries of a Postlabor World

Factory robots of today are unrecognizable from the image of the production line during the heyday of the US automotive industry (1908–73). The massive steel mechanical arms filling rooms longer than a city block, each performing a single task in the construction of a vehicle with repeated mechanical precision, have been replaced by robots that are decidedly "smart." The interactive robots of the present, able to perform and respond to human and machine social cues, seem to be a different species from the robotic arms that posed a physical danger to the comparatively vulnerable human bodies that labored alongside them. Designed with artificial intelligence and "deep-learning technology" so that they can observe and correct their own errors and those of robots with which they work, these robots, which are used in production plants to assemble iPhones, Volkswagens, and Tesla electric and self-driving cars, can reprogram themselves to optimize performance and take on new tasks during the hours human workers sleep.[1] Because automated teaching and learning characterizes this new generation of factory robots, they are seen to take the human "out of the loop." This is a central aspect of what the World Economic Forum has recently dubbed the "Fourth Industrial Revolution," in which even the production and programming of technology can be performed by machines themselves.

Human-free industrial robotic production seems to announce that an ostensibly postlabor (and as we will go on to argue in this chapter, a "post-race" and "postgender") epoch is upon us. Since the first industrial revolution, automation has signaled the threat of the replaceability of specific types of human functions and human workers that are racialized and gendered. Following the 2016 US presidential election, for instance, the media announced that Donald Trump could not live up to his promises to return jobs to the white working class because these jobs were lost to robots rather than threatening racial others (that is, "illegal" immigrants and racialized outsourced labor in the Global South). Nonetheless, the machine and the racialized other pose a similar threat of displacing those already fully human in this conception of white loss.

This chapter assesses the shifting contours of US racial liberalism and white supremacy through readings of philosophical, cultural, media, and political accounts of US labor and automation in the twentieth and early twenty-first centuries. We excavate the suppressed racial imaginary that produces the subjects and objects of present-day "technoliberalism" as a political form that elaborates the logics of liberalism and reinvigorates the imbrications of liberalism and fascism at the heart of the US formation as a racial state. By technoliberalism, we mean the ideology that technology advances human freedom and postracial futurity by asserting a postlabor world in which racial difference, along with all human social difference, is transcended.

Yet, as we argue, what is essential about the automation of both factory and domestic labor for technoliberalism is its production of the surrogate human effect—that is, a racial and gendered relation emerging at the interstices of new technologies and the reconfigurings of US geopolitical dominance. By displacing the centrality of racialized and gendered labor relations in its articulation of a postracial present enabled by automation and artificial intelligence, technoliberalism reproduces the violent racial logics that it claims to resolve through techno-scientific innovation. In this sense, technoliberalism is an update on the liberal progress narrative that conceals ongoing conditions of racial subjugation and imperial expropriation.

Racial liberalism has long been viewed as a central tenet of governance and empire in the United States. Jodi Melamed has argued that,

In contrast to white supremacy, the liberal race paradigm recognizes racial inequality as a problem, and it secures a liberal symbolic framework for race reform centered in abstract equality, market individualism, and inclusive civic nationalism. Antiracism becomes a nationally recognized social value and, for the first time, gets absorbed into U.S. governmentality. Importantly, postwar liberal racial formation sutures an "official" antiracism to U.S. nationalism, itself bearing the agency for transnational capitalism.[2]

The technoliberal form espouses antiracism as a central premise even as it affirms the racial order structuring technological modernity. As the grammar of technoliberalism, the surrogate human relation innovates the logics of racial hierarchy by producing a fantasy of a postracial world enabled through technology that functions as service. Put otherwise, technoliberalism's postracial imaginary produces human freedom as an effect of a postlabor world, which is an updated variation of the racial liberalism that has been a part of US governance from the inception of the nation. By insisting on the continuing importance of racial and gendered histories of labor that expand the category of labor and the worker far beyond the myth of the factory worker as the only subject of labor, we can see that technoliberalism not only fails to be antiracist, but is in fact an affirmation of US racial logics governing what forms of labor are visible as such, and what forms are not. Because of the technoliberal assertion that we are entering a postlabor world that is implicitly anti-racist because it is human-free, it is urgent to assess the ways in which technology has innovated upon capital's dependence on racialized and gendered labor.[3]

In order to track the relation between the politics of technological futures and figurations of human freedom, we foreground how cultural ideals about human value and potential have been shaped through the fantasy that degraded tasks, that is, those that are are racialized and gendered, could be fully automated. To do so, this chapter first considers the ways in which these fantasies are the inheritors in form and function of enslaved and indentured labor. Though modest in the sense that we do not attempt a comprehensive or encyclopedic history, this chapter allows us to see how automation fantasies (1) depend upon slavery and the idea of a worker who cannot rebel and (2) are contiguous with existing labor exploitation along colonial/racial lines. This history provides a foundation for apprehending how the surrogate effect produces freedom as a racial form. Next, the chapter turns to Cold War, and then post–Cold War developments in the political terrain of technoliberalism. We survey the relationship between the discourse of "white loss," immigration, and automation in contemporary US national politics, and evaluate technoliberal imaginaries that argue that it is robots, not racialized others, who are taking US jobs. As we argue, such imaginaries pin their antiracist logics on a postlabor future. Throughout the chapter, we thus reflect on the historical links between the seemingly opposed logics of white supremacy and racial liberalism.

Automation and the Worker without Consciousness

The fear that automation heralds human obsolescence is in one sense as old as the modern systems of labor and production. At the same time, human obsolescence is also always in and of a dystopic future not quite yet upon us. Starting with the Luddites, English workers who destroyed textile mills in a series of riots beginning in 1811 because of fears that they would lose their jobs, there have been waves of response to the development of industrial technologies that have all led to the same question: When will human labor be rendered completely redundant by machine labor? In 1930, the economist John Maynard Keynes coined the term *technological unemployment* to capture a dilemma similar to that of the Luddites, namely the problem that escalates when technological developments outpace society's ability to find new uses for human labor. At the same time, Keynes argued that this was only a temporary phase of maladjustment.

The specter of human obsolescence undergirds both utopic and dystopic approaches to technological modernity and the surrogate effect of technologies that are hoped to liberate, and at the same time feared to obliterate, the human subject. Automation and technological futures make especially legible the ways that labor continues to be a central site through which freedom and unfreedom racially demarcate the bounds of the fully human. If the human condition (in the Arendtian sense) is defined in terms of how one acts in and upon the world, including through work and labor, then dreams of mechanization are never just about efficiency, but also inevitably about the kinds of work and labor that are unfit for a human to perform.[4] In this way, we can think of automation as historically tied to the promise and threat of the liberation of human laborers. Put otherwise, even as machines enact the surrogate human effect in areas that can be automated, they also produce a surrogate (nonworker) humanity liberated from prior states of unfreedom. Yet automated utopias in which the human worker has been emancipated from miserable (dull, degrading, and repetitive) work are also tethered to dystopias in which the liberated subject is dissolved as replaceable and, therefore, potentially obsolete.

Hannah Arendt argued that labor, work, and politics were central to how the tensions between human freedom and unfreedom would develop as automation played an ever-increasing role in social and economic structures. In the 1958 preface to *The Human Condition*, Arendt speculates that automation, "which in a few decades probably will empty factories and liberate mankind from its oldest and most natural burden, the burden of laboring and the bondage to necessity," will lead to a society that no longer knows its purpose.[5] As she elaborates, "It is a society of laborers which is about to be liberated from the fetters of labor, and this society does no longer know of those higher and more meaningful activities for the sake of which this freedom would deserve to be won. . . . What we are confronted with is the prospect of a society of laborers without labor, that is, without the only activity left to them."[6] Arendt outlines the *vita activa*, a philosophy detailing three fundamental human activities: labor, work, and action. For Arendt, labor is what must be done to reproduce human life; work creates the world of things; and action, the third human condition, constitutes our social being, and is therefore inherently political. She proposes that as workers entered society and were no longer outside of it (after the granting of an annual wage that transformed a "class society" into a "mass society"), the potential of the radical political movement as a workers' movement dissipated.[7] Arendt's fears grew out of her contention that we have become a society where to be a subject means being a laborer, but without the step of proletarianization that was the engine of Marx's teleology of anticapitalist revolution. Proletarianization was the essence of Marx's new historical subject of labor, which arose out of industrialization and the need to sell labor for a wage. Arendt was concerned that in distinction to what Marxists would believe, without the unifying class experience of laboring for a wage in a laborless society, and without the social and interactive condition of what she calls the vita activa, the promise of technological liberation is empty.

Crucially, Arendt's figure of the laborer as social-historical subject is based upon her distinction between slavery as a nonmodern economic form and modern capitalism. She writes, "In contrast to ancient slave emancipations, where as a rule the slave ceased to be a laborer when he ceased to be a slave, and where, therefore, slavery remained the social condition of laboring no matter how many slaves were emancipated, the modern emancipation of labor was intended to elevate the laboring activity itself, and this was achieved long before the laborer as a person was granted personal and civil rights." Arendt writes that "the incapacity of the *animal laborans* for distinction and hence for action and speech seems to be confirmed by the striking absence of serious slave rebellions in ancient and modern times."[8] She distinguishes this from the successes and revolutionary potential of the 1848 European revolutions and the Hungarian revolution of 1956, organized around the working class.

According to Nikhil Pal Singh, the tendency to separate slavery (or the state of unfreedom) from the period of modern capitalist development misses the ways in which "the chattel slave was a new kind of laboring being and a new species of property born with capitalism."[9] The dependence of chattel slavery on race concepts in turn fostered "the material, ideological, and affective infrastructures of appropriation and dispossession that have been indispensable to the rise of capitalism."[10] According to Singh, not only does the "differentiation between slavery and capitalism [widen] the gulf between slaves and workers," but it also "overlooks how racial, ethnic, and gender hierarchies in laboring populations are retained as a mechanism of labor discipline and surplus appropriation, and even as a measure of capitalism's progressivism, insofar as it purports to render such distinctions anachronistic in the long run."[11] For Arendt, the plurality of a differentiated polity is the condition of possibility for freedom. Thus, her collapsing of Nazi fascism and Soviet totalitarianism as equivalent conditions of unfreedom for the undifferentiated masses affirms a liberal politics. Instead, Singh proposes that "Capital ceases to be capital without the ongoing differentiation of free labor and slavery, waged labor and unpaid labor as a general creative basis. . . . Only by understanding the indebtedness of freedom to slavery, and the entanglement and coconstitution of the two, can we attain a critical perspective adequate to a genuinely anticapitalist politics."[12] Singh's critique points not only to the ongoing imbrications of conditions of unfreedom and liberal capitalist development, but also to how the disavowal of the worker without consciousness (the slave, the unfree) depends upon the false dichotomy between fascism and liberalism.

This elision of the entanglement of fascism, racial violence, and liberal developmentalism is replicated in the technoliberal imaginary of human obsolescence in a postlabor world. The racial and gendered structures of production, both material and social, that continue to demand an abject and totally submissive workforce reevidence themselves in the practices and fantasies surrounding the role of robot workers. Utopic hopes of freeing human workers through evolving iterations of unfree labor (as automated solutions to the racial contradictions upon which free labor is predicated) also bring forward colonial fantasies of the other as a nonsubject. Lacking political consciousness, the robot, which stands in for colonized and enslaved labor, cannot stage a revolution. In this sense, the freedom of the fully human liberal subject cannot come to be without the unfreedom of the less than human or the nonhuman.

As is well documented, the first use of the term *robot* appeared in Karel Čapek's 1920 play R.U.R. (Rossum's Universal Robots) and is derived from the Czech word for slave. We also find here the first instance of a human being defined as that which has mastery over the robot, and a robot as that which obeys the human. The play begins in a factory that produces robots, artificial people who are cheap and readily available. The robots make commodities for a fraction of the cost that human beings could. Amid debates among the characters about whether robots have souls and deserve rights and liberation, or whether they are appliances, there is a widespread robot rebellion that wipes out human beings. The fear of the robots' revolution is caught up in debates about the "humanity" of robots (importantly hinging on whether or not they have souls) and the potential end of human civilization.

This literary history of the robot slave can productively be put into conversation with the modern history of liberal ideas about human freedom, which are entangled with the abolition of the transatlantic slave trade and the emergence of free labor as a racial category.[13] A body meant for use and toil is viewed as being without a soul and endlessly exploitable. The idea that some bodies are meant solely for work informs fantasies about automation. These are the discourses of European territorial colonialism and chattel slavery, revivified as the human (white, European, liberal, male) that now must be distinguished from the machine and preserved in the face of its otherness. The question of whether or not the colonial other, embodied in discourses of the primitive and the native, had a soul were the justification for the evacuation of humanness that enabled the conquest, genocide, and enslavement of European territorial colonialism. Saving the soul of the racial other, meanwhile, implied that such an other could be put on a trajectory toward freedom in a way that mechanical workers cannot. As Despina Kakoudaki argues, there is a direct representational relationship between the racial histories of human chattel slavery and the robot's "potential for racial or ethnic representation [that] comes from its objecthood. . . . Robots are designed to be servants, workers, or slaves" since their "ontological state maps perfectly with a political state."[14]

Imaginaries that servitude could be mechanized, thus eliminating the need for racialized bodies (whose function within modernity was always already to serve), extend into the present moment of rethinking human labor through new robotic technologies. Kakoudaki insists that "the fantasy of the robotic servant, worker, or slave promises that if enslavement of real people can no longer be tolerated in the modern world, then mechanical people may be designed to take their place, and their labor will deliver the comforts of a laborless world for the rest of us."[15] In this way, we can see robot workers as following the pattern established by British imperial labor allocation, in which the abolition of slavery in 1809 was enabled by the simultaneous instantiation of indentured servitude to replicate slave labor, now "free."[16] Robots, like indentured workers, function as a category that produces a false "freedom" that holds open the place of possibility for the reproduction of the (white) liberal subject in the absence of older forms of unfree and coerced labor.

One particularly evocative example is Westinghouse Electric Corporation's novelty "mechanical man," Rastus the Mechanical Negro, unveiled for the electrical and radio exhibitions of 1930 (figure 1.1). A 1931 *Radio Craft* article refers to Rastus as a "mechanical servant" and "robot," and marvels that "He moves (stands up or sits down, bows), talks (6 word vocabulary or sequences), and [has] a photo-electric cell in his head, [which] when activated by a light, in this case a beam from the arrow, is triggered, [and] a gunpowder charge blows the apple off his head."[17] The author of this piece is impressed with Rastus's range of motion, particularly that he is able to bow, although as the author notes, Rastus has not yet mastered walking. Rastus's usefulness as a mechanical servant is thus limited, even as the mechanization of his racialized gesture of obeisance causes delight for the journalist. Rastus, which another article refers to as a "mechanical slave," is also distinguished by being a target for humans wishing to use an electrical beam shot from an arrow. Thus, in addition to obeisance, Rastus is a racial object toward which violence (in spite of the whimsical reference to William Tell's marksmanship) can be unleashed.

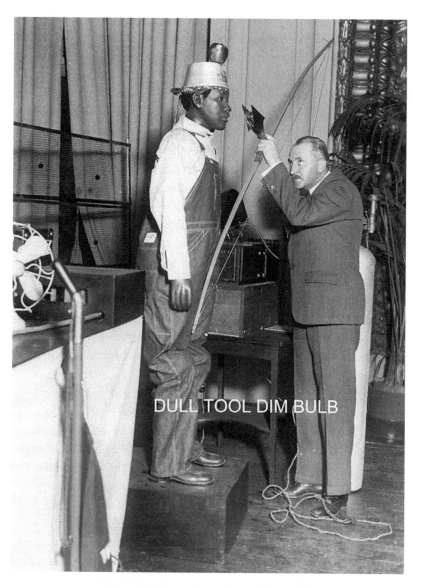

Figure 1.1. Rastus the mechanical "servant."

As Rastus exemplifies, robots have historically embodied "a form of abjection that is total and absolute, a state of preconsciousness in which neither the gestures of emancipation nor those of revolt are available."[18] The total submission of the robot to the human as the robot-slave enables a formulation of human freedom from labor without submitting the human to the machine. Moreover, the robot worker, who replaces the slave in the twentieth century, is one who can never come to political consciousness or question the economic or political system within which it toils. Unlike the racialized human worker, the robot-worker is in a perpetual state of false consciousness, unable to rebel, go on strike, or unionize because robots are supposed to "have no fantasies about their social position."[19] Because robots introduce a form of worker who does not threaten to revolt, it is no coincidence that they emerge as a predominant cultural figure for labor in the age of industrialization. This was a time when the nature of work was being dramatically refigured through automation, including Fordism and Taylorism, with a focus in maximizing productivity. At the same time, the 1917 Russian Revolution and the first "red scare" also inflected the hopes and fears tied to the worker in the age of mass production.[20] Communist revolution stood as a limit to the amount of exploitation workers would (or could) be willing to undergo. The worker who could not rebel and was always productive—the robot—thus became an important trope for emergent fantasies of capitalist development unfettered by threatening, organized human workers.

The connection between the category of the human and mastery over the technological means of production continued to be important throughout the twentieth century. In a 1943 editorial urging black Americans to support the Allied war effort against fascism while being vigilant of white supremacy, Ralph Ellison argued that "Negro leadership" must master "defining the relation between the increasing innovations in technology and the Negro people's political and economic survival and advancement. During the war the mastery of hitherto unavailable techniques by Negroes is equivalent to the winning of a major military objective."[21] This mastery, Ellison suggests, will be crucial for effectively leading working-class struggles and freedom, which must be won and not gifted. The story of technological mastery is positioned in Ellison's text in opposition to the John Henry myth—the myth of the black steel driver, a physically laborious job necessary for expanding the railway that involved driving steel into rocks to make holes for explosives. In the story, Henry races a steam-powered hammer and wins, only to die when his heart gives out from overexertion after achieving victory. Ellison is suggesting that rather than being the tools (objects, servants, endlessly exploitable for increased production), black Americans must become the masters of technological objects if they are to become free.

At the same time that robots like Rastus reaffirmed the racial figure of the human against the abject robot slave, and Ellison's 1943 plea for black Americans' support of the US war effort affirmed that racial uplift could be attained through technological mastery, numerous twentieth-century cultural texts grappled with the fact that automation had blurred the boundary between the human and the machine. These texts can be read as a reflection on how capitalism must continually redefine the bounds of the human, as well as on how the promises of freedom offered by automation are continually haunted by the states of unfreedom undergirding racial capitalism. For instance, Fritz Lang's film *Metropolis* (1927) not only portrays the awe and horror of the newly electrified city, but also insists on examining what lies beneath industrial lights and train skyways. As the protagonist Freder asks his industrialist father toward the beginning of the film, "Where are the people whose hands built your city?" They are, in fact, relegated to Workers' City, which is located underground (in the depths underlying the Metropolis). Out of sight, they toil and are used up by "the machine" that controls the marvels of urban modernity. The film culminates in a worker rebellion instigated by a robot, who pretends to be one of the leaders of the workers. The film thus connects two concerns: the literal subjugation of the worker to the machine, symbolized in the underground area of the Metropolis, and the human becoming the machine, symbolized in the robot who looks exactly like her human counterpart.

Techno-dystopic fantasies revolve around the reversal of normative presumptions about the social need for a totally abject and submissive workforce. Cultural fears that those enslaved as objects rebel—even when they aren't supposed to—abound, as is evident in texts such as *Metropolis*, Karel Čapek's R.U.R., or the television series remake of *Battlestar Galactica*, in which human identical cybernetic aliens ("Cylons") infiltrate society with the goal of eradicating human civilization. The subsumption of the human into the machine, and vice versa, was theorized by Marx in volume 1 of *Capital*. He writes of workmen "as merely conscious organs, coordinate with the unconscious organs of the automation," such that automation is the subject and workmen merely its object.[22] Thus, "the machine possesses skill and strength in the place of the worker. . . . The worker's activity, reduced to a mere abstraction of activity, is determined and regulated on all sides by the movement of machinery."[23] These fears, particularly in the twentieth century, are tied not to actual robots, but rather to the human condition within industrial capitalism.[24] Put otherwise, the condition of being human is already tethered to the human's coproduction with the machine as something whose parts (and whole) are alienable.

Cold War Automation

As we've argued thus far, in the US, robot–human labor relations and attendant notions of freedom and unfreedom have been determined by the structures of racial capitalism. Automation and mechanical replacements for human laborers function much like the figure of the emancipated worker who stands in seeming opposition to the unfree, enslaved worker in traditional Marxist thought in the sense that they engender a temporal trajectory to capitalism—one that ends with human emancipation from toil and oppression. At the same time, as we argue here, it is crucial to attend to precisely those moments where the promises of technological modernity in fact reiterate and reaffirm the racial-colonial structures of unfreedom that support and maintain the autonomous liberal subject. While the early-twentieth-century texts we analyze explicitly position automated technologies as the inheritors of the enslaved (racial) body, post–Cold War texts contextualized in the early civil rights era posit that automation will fully emancipate the formerly enslaved. However, the threat of human replacement also produces a dystopics of the dissolution of the liberal subject. Automation thus also always foments symbolic and (at times) physical violence against racialized workers in the face of joblessness caused by automation.

In a representative 1964 episode of *The Twilight Zone*, "The Brain Center at Whipple's," Whipple is a factory owner determined to replace men with machines (figure 1.2). The series' host Rod Serling's opening voiceover announces the theme: "These are the players, with or without a scorecard: In one corner, a machine; in the other, one Wallace V. Whipple, man. . . . It happens to be the historical battle between flesh and steel, between the brain of man and the product of man's brain. . . . For this particular contest there is standing room only in . . . the twilight zone." One by one, the posi-

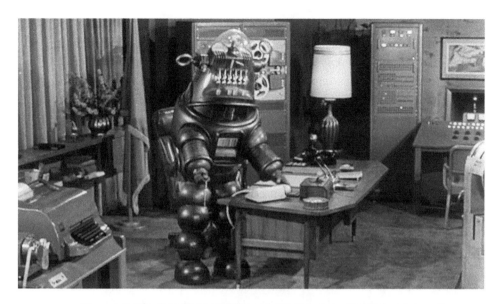

Figure 1.2. "The Brain Center at Whipple's," *Twilight Zone* (Season 5, 1964).
A robot replaces the factory owner himself.

tions in Whipple's plant are robotized, and various workers are replaced, starting with the men on the line, then the foreman, and, eventually, even the engineers who service the machines. The men accuse Whipple of trading in the pride that a single man takes in his work for mechanization. Yet if the mourning of the disappearance of the skilled craftsman who finds his identity in the work he does with his hands was a familiar narrative by the mid-1960s, the episode's final twist was, perhaps, less so. In the end, a robot replaces Whipple himself, the factory owner and cost-cutting, downsizing capitalist par excellence. This ending alludes to the eerie possibility that mechanization might bring about the end of capitalism itself.

Since the end of World War II, the desire to eliminate the worker as the subject of revolution went hand in hand with the perpetuation of capitalism, innovating upon the imperial-racial formulations of freedom as free wage labor. With the advent of the Cold War and the acceleration of automation, freedom also came to designate the distinct spheres of Soviet and capitalist lifeworlds, which was reflected in how automation articulated with shifting notions of technology and politics. Both communist and capitalist lifeworlds are of the Enlightenment, and both therefore necessarily contended with mechanization and automation as part of a radical break with the past.[25] Yet, because mechanization, automation, and industrialization structure both utopic and dystopic imaginaries of human–machine entanglement, how cultural and political discourses framed automation became crucial in distinguishing capitalist from communist technological advances in the field of labor. This distinction was central to the establishment of US Cold War imperial dominance. As Singh argues, Cold War discourses around totalitarianism that determined attendant notions of freedom (located in the US and Western Europe) "not only linked fascist destruction of what Arendt termed 'authentic political life' to the Soviet Regime, [they] also suggested an extended chain of reasoning about existential dangers posed by 'terrorist uses' of technology by those lacking proper philosophical conditioning and historical preparation for exercising state power."[26] As fascism was excised from the realm of the West to that of the East, then, certain modes of automation, especially those that reduced the human to the machine, came to be associated with ongoing states of unfreedom justifying US "humanizing" imperial violence in the decolonizing world where the Cold War was fought.

On the one hand, then, the replacement of human workers with robots, especially in the early years of the Cold War, came to be correlated with the mechanization of totalitarianism. In 1951, the *New York Times* reported that the Soviets had startlingly created an all-robot factory, "where work is done by machines, not men."[27] Complete mechanization was also at times billed as a Soviet dream. For instance, an article covering the Soviet proposal for automating its railways explained that the Soviet deputy premier held firm to the belief that a robot "would be more efficient than the most experienced human engineer."[28] Such automation projects were associated with the Soviet push toward rapid industrialization, which did not care about human welfare, and, as the article concludes, with an inhumane pace of change; whereas before the revolution, what became the Soviet Union produced only "2 or 3 percent" of "the world's total industrial output," by 1958 it was producing one-fifth of that output.[29]

Accounts of Soviet automation were foundational to articulations of human freedom in the US in the face of totalitarian and fascist states of human unfreedom. As Scott Selisker argues, figurations of the human who has been reduced to a machine (through brainwashing or through the horrors of technological modernity) stands as the preeminent threat that totalitarianism poses to US liberal individualism and exceptional freedom.[30] US Cold War racial liberalism can in this sense be viewed as central to the project of celebrating US diversity in contrast to the undifferentiated mechanized masses of the Soviet Union. This is not to say that techno-logical innovation was not celebrated in the US Cold War context. Instead, technological innovation was celebrated in such instances when, rather than reducing the human to the machine, the machine freed the human for intellectual and creative capacities. For instance, in 1962, when the Soviets proposed instituting a driverless subway train, news coverage pointed out that it was to be modeled on a three-car robot train that ran between Times Square and Grand Central Station, thus implying that American ingenu-ity had beaten the Soviets to the punch.[31] In 1955, when Cuthbert Hurd, a division director for IBM, proposed that the "completely automatic fac-tory" was a near possibility, he also emphasized that this would "not make man obsolete."[32] Rather, as he explained, it would allow man true freedom, including time for the "'higher activities of the mind."[33] Automation could thus be positioned as yet another step in freeing labor on the path toward US democratization following emancipation. But it was not just engineers who pinned high hopes to the improvement of human life through automa-tion. For example, union chief Walter Reuther criticized President Eisen-hower's 1955 economic report for failing to take into account how the likely automation of industry and commerce in the next decade or two could be used for future economic growth and the improvement of general living conditions for the American people.[34]

One of the first of the early industrial robots, the Unimate, a 4,000-pound arm attached to a steel base, was put to use in a General Motors factory in 1961 (figure 1.3). Unimate robots "could easily pour liquid metal into die casts, weld auto bodies together and manipulate 500-pound (227-kilogram) payloads."[35] Mechanical arms based on hydraulic actuators not only were exceptionally precise in their tasks, but they dramatically increased productivity. Thus, for instance, "Australia's Drake Trailers in-stalled a single welding robot on its production line and benefited from a reported 60 percent increase in productivity."[36] Doing repetitive tasks with a simple program, industrial robots like the Unimate did carry out the threat of eliminating factory jobs that were considered unskilled. On the whole, however, news coverage about the US technological ascendancy emphasized that automation leads to improved material conditions—and even potentially to overcoming prior states of unfreedom.

Cold War accounts of automation as the condition of possibility for social and economic betterment reinscribed (even as they redescribed) histories

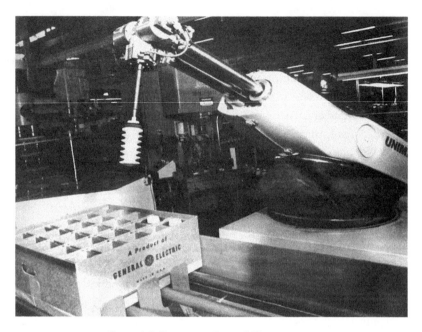

Figure 1.3. Unimate at General Electric, 1963.

that produce freedom for the liberal subject at the expense of the unfreedom of raced and gendered workers in US history. We can say that the acceleration of automation during the Cold War also marks the beginning of a new version of the liberatory imagination in which unfree (racialized) workers are simultaneously erased yet necessary for the production of the myth of racial uplift. For example, in 1952, the *New York Times* ran a story on factory automation in the US South. H. M. Conway, the director of the Southern Association of Science and Industry, refuted the criticism that "the South was designed to exploit cheap labor 'with rows of machines operated by workers only slightly removed from slavery.'" Rather, he argued that mechanization was about labor-saving "American ingenuity."[37] The report concludes with the example of a synthetic fiber plant in Alabama, in which it claims just a few hundred workers could turn out millions of pounds of fiber. This example implies that it is not the case that black workers are still exploitable in the US South, but rather that technology eliminates the need for a population of racialized, exploitable workers in advance of the production process.

The eradication of racialized (and enslaved) labor through automation illustrated in this example about the 1950s US South echoes the racial logic and fascist discipline enacted by techno-utopianism in California's agri-businesses, as tracked by Curtis Marez in *Farm Worker Futurism*.[38] Marez argues that "for much of the twentieth century . . . agribusiness futurism [projected] a future utopia in which technology precludes labor conflict and where new machines and biotechnical research [could] eliminate some workers while enabling the expanded exploitation of others. Agribusiness futurism fetishizes new technology in an effort to eliminate resistance and subordinate workers of color to the machinery of production."[39] Thus, even as speculative figurations of future technologies like robot plows and mechanical farmworkers, and actually existing technologies like mechanical orange pickers, projected a fantasy of eliminating the need for racialized labor, in fact, they also provided new ways for agribusiness to reorganize and control an expanded workforce.[40] Because certain tasks were deskilled and downgraded due to automation, the use of technology widened the contradictions between the motivations of capital versus labor, as evidenced by the Bracero program.[41]

Cold War accounts of automation provide a historical account of the racial calculus of the surrogate human effect. Race thus haunts even those imaginaries that fully remove the human from the frame of representation. One such example is the 1982 short film *Ballet Robotique*, which is the only industrial film to have been nominated for an Academy Award. The film paid homage to changes in the automotive industry. The film re-imagines the auto plant as the site for a choreographed, elegant, and most importantly, human-free dance among robots.[42] *Ballet Robotique* depicts the movements of huge assembly-line robots, which are synchronized to London's Royal Philharmonic Orchestra's rendition of classical music that includes the Habanera from Bizet's opera *Carmen* and Tchaikovsky's *1812 Overture* (figure 1.4). Showers of welding sparks punctuate the dance, and the film has more the feel of a space-age science fiction movie than an industrial film. "The title 'Ballet Robotique' echoes Fernand Leger's classic avant-garde film 'Ballet Mechanique,' from 1924. . . . Each robot was programmed in a different way, suggesting an interaction of multiple robotic personalities. Borrowing from animal and plant motion, the filmmakers had the robots programmed as a chicken, a cat, and a swaying and curling seaweed."[43]

Figure 1.4. A robotic ballet: Human-free elegance.

In spite of the musical and choreographed attempts at showing automa-tion in the auto industry as part of the development of Western civilization, art, and culture, the film strikes a chilling note when one recalls that the 1980s were a period of unprecedented crisis for the US auto industry and its employees. "After peaking at a record 12.87 million units in 1978, sales of American-made cars fell to 6.95 million in 1982, as imports increased their share of the U.S. market from 17.7 percent to 27.9 percent. In 1980 Japan became the world's leading auto producer, a position it continues to hold."[44] This shift was in part due to the inefficiency of US cars in the midst of the oil crisis, and in part because of the overall superior quality of Japa-nese cars. The automation portrayed in *Ballet Robotique*, which included the increased integration of industrial robots on the assembly line, was part of an attempt by US automakers to retake their share of the market. GM, Ford, and Chrysler engaged in major layoffs to achieve higher profits and lower overhead costs.

In spite of the ostensible successes of Cold War civil rights and emergent articulations of racial liberalism, because racial violence is the condition of possibility for racial capitalism, it is not surprising that each encounter with what appears to be a new moment of human obsolescence enacted through automation becomes easily associated with white loss. Layoffs in the US auto industry, perceived by the public as a response to the dominance of Japanese automobile manufacturing, led to a series of attacks on Asian Americans. This includes the infamous 1982 beating of Vincent Chin in Detroit by two laid-off Chrysler workers, which resulted in Chin's death. Though it was presented by the defendants in court as a bar brawl gone bad, eyewitnesses reported that the killers justified their attack by saying, "It's because of you we're out of work," linking the Asian American Chin to Japanese autoworkers.[45]

The conflation of productivity and robotic function with Asian-ness was not new to the 1980s. As Wendy Chun has argued, "the human is constantly created through the jettisoning of the Asian/Asian American other as robotic, as machinelike and not quite human, as not quite lived."[46] Yet the 1980s did signal a shift in Asia's position within the global economy and the refiguration of the human–machine boundary in the late years of the Cold War and the early decades of the postindustrial era. According to David Roh, Betsy Huang, and Greta Niu, "the yellow peril anxiety of an earlier, industrial-age era embodied by Fu Manchu found new forms across cultures and hemispheres as Asian economies became more visible competitors."[47] They propose techno-Orientalism as a concept through which to apprehend the ways in which racial, technological, and geopolitical imaginaries are linked in order to secure US dominance.

> Techno-Orientalist imaginations are infused with the languages and codes of the technological and the futuristic. These developed alongside industrial advances in the West and have become part of the West's project of securing dominance as architects of the future, a project that requires configurations of the East as the very technology with which to shape it. Techno-Orientalist speculations of an Asianized future have become ever more prevalent in the wake of neoliberal trade policies that enabled greater flow of information and capital between the East and the West.[48]

As the authors elaborate, part of these changes involved the differentiation of China and Japan. Whereas Japan was threatening to the US as a place that also innovated technology, China was threatening as a space with which the US competed for labor and production—"Japan creates technology, but China *is* the technology," and a human factory.[49] In both instances, however, techno-Orientalism portrays the inhumanity of Asian labor in the face of Western liberal humanism.[50] Thus, the Asian body has been produced as an "expendable technology" at the heart of US industrialization.[51] In this context, they provide the following reading of Vincent Chin's murder:

> The callous brutality of Chin's death evinces something more than racial hatred; Chin not only was perceived as a convenient stand-in for the Japanese automotive industry, but embodied its traits— unfeeling, efficient, and inhuman. In Ebens and Nitz's eyes, they were Luddites striking down the automatons that had been sent in to replace them. Techno-Orientalist discourse completed the project of dehumanizing Vincent Chin by rendering him as not only a racial-ized Other, but a factory machine that had to be dismantled by Ebens and Nitz to reclaim their personhood, subjectivity, and masculinity.[52]

Ballet Robotique, when contextualized through the disenchantment of white factory workers in the 1980s, signals that the automation of the postindustrial era threatened the very structures of white supremacy that had been upheld by factory automation in the early years of the Cold War. In this sense, both the early and late Cold War figurations of human obso-lescence and human freedom, through the racial grammar of the surrogate human effect of robotic arms, driverless trains, and racialized outsourced laborers who function as stand-ins for technology, inherit and reiterate local and global racial forms. The racial violence that inheres in the ma-chines that promise freedom for those (already) fully human and threaten obsolescence (as white loss) continues to structure present-day imaginar-ies of technological development and automation. Yet, as we argue in the next section, the linkages between fascism, racial violence, and technology as the conditions of possibility for the thriving of the autonomous liberal subject have been once again covered over in narratives surrounding capi-talist development toward human freedom. This is the emergent political formation of technoliberalism.

White Loss and the Logics of Technoliberalism

One of US President Donald Trump's first executive actions was the pro-nouncement on January 24, 2017, that he would begin immediate con-struction of a border wall between the US and Mexico. Both Trump and the phantasm of a border wall promoted in his presidential campaign are symbols that bring together labor, technologies of control, and white supremacy. The fantasy of isolation, closing of borders, and expelling immigrant labor animating Trump's promise to build a border wall was about reviving a formerly booming national economy. The border wall, in short, is about restoring the (unmarked white) working-class laborer to a former glory, symbolized by figures like the auto industry worker and coal miner.

In response to claims that immigrant others are responsible for the loss of a glorious era of white labor in the US, following Trump's election, tech industry leaders emphasized that automation, not immigration, was the biggest threat to American workers. These claims were positioned as anti-racist, redirecting concerns about white loss from the immigrant threat to the threat of robotics. Yet, as we contend, these arguments are also part of a problematic relationship between technoliberalism and the fantasy of postracism in the labor arena. It is thus crucial to ask: What is the relationship between white loss and the discourse of postlabor technology/technoliberalism? On the one hand, Trump's campaign and presidential rhetoric, which has been overtly racist in its tenor, contradicts the tenets of racial liberalism in US politics and culture outlined by Melamed. On the other hand, we might consider the opposition by tech leaders to Trump's assessments of job loss as white loss as yet another innovation in the logics of racial liberalism—what we are theorizing as the emergent formation of technoliberalism that gains force in the post–Cold War era.

By technoliberalism, we mean the way in which liberal modernity's simultaneous and contradictory obsession with race and the overcoming of racism has once again been updated and revised since the dissolution of the threat of Soviet totalitarianism through the frame of a postlabor world. Technoliberal futurity envisions robots as replacing human form and function in factories and other spheres (for instance, bus and taxi drivers, janitors, and call center workers). As proponents of such futures argue in seeming opposition to the overtly racist and fascist logics of Trump, it is not that jobs are being stolen from the US white working class by Mexicans and the Chinese, but that, as robots take human jobs, other solutions for human obsolescence must be found. Because technoliberal explanations of economic transformation utilize the vocabulary of color-blindness, there is a need to map out how race scaffolds the technoliberal imaginary and its seeming opposition to white supremacy.

One representative example is a CNBC commentary that noted that "Eighty-five percent of manufacturing jobs were lost to technology from 2000–2010, not to Chinese job thieves. . . . We can expect more automation, and not just on the shop floor. Last year the White House predicted 47 percent of existing jobs will eventually be eliminated by automation and artificial intelligence. . . . Unless Donald Trump has a time machine stored in a secret lair underneath Trump Tower, going back to a labor-driven manufacturing economy isn't an option."[53] In the technoliberal logic of this report, it is not that the new economy uses cheap, racialized labor outsourced to the Global South. Rather, *the new economy doesn't need any human labor.* Similar assessments abound in news reports pointing to Trump's racist fallacies, including his cornerstone promise to revive coal mining in the US (and thus dismantle the Obama-era clean power plan). Yet, as a recent *New York Times* article outlines,

> In Decatur, Ill., far from the coal mines of Appalachia, Caterpillar engineers are working on the future of mining: mammoth haul trucks that drive themselves. The trucks have no drivers, not even remote operators. Instead, the 850,000-pound vehicles rely on self-driving technology, the latest in an increasingly autonomous line of trucks and drills that are removing some of the human element from digging for coal. . . . Hardy miners in mazelike tunnels with picks and shovels [celebrated by Trump]—have steadily become vestiges of the past.[54]

What is interesting about this article and others like it is that the evidence that Trump's labor policies are racist is the same evidence that suggests that the laborer as a figure, and human labor writ large, is vestigial to the tech-driven economy. To be clear, we are not contending that the nostalgic vision of the white working class in the US is disconnected from the logics of white supremacy. Rather, we are proposing that race continues to structure technoliberal futures even as they claim to be postracial.

For instance, the fully automated future certainly is not unconditionally celebrated within technoliberalism; it is often articulated as loss—however, not as specifically white loss, but rather as universal human loss. According to Stephen Hawking, "The automation of factories has already decimated jobs in traditional manufacturing, and the rise of artificial intelligence is likely to extend this job destruction deep into the middle classes, with only the most caring, creative or supervisory roles remaining."[55] Elon

Musk, founder of Tesla and SpaceX, makes a similar point: "What to do about mass unemployment? This is going to be a massive social challenge. There will be fewer and fewer jobs that a robot cannot do better [than a human]. These are not things that I wish will happen. These are simply things that I think probably will happen."[56] Mark Cuban, the tech entrepreneur and celebrity judge on the popular ABC television series *Shark Tank*, has been critical of Trump, and in particular of Trump's racist understandings of job loss. As he put it, companies that promised to build new factories and create new jobs in the US, such as Fiat Chrysler, GM, Ford, Intel, and Walmart, "when it is all said and done . . . are going to end up leading to fewer people being employed . . . and there is nothing President Trump can do to stop that because of the trends in technology—machine learning, neural networks, deep learning, etc.—changing the nature of work."[57]

At the same time that tech billionaires articulate the end of the human worker as a universal human loss against Trump's anti-immigration stance, the connection between liberalism and capitalist development means that warnings about the inevitable obsolescence of manual jobs is paired with a simultaneously optimistic outlook on the growth of the capitalist economy. Thus, the technoliberal imaginary continually reasserts that US industry is growing, rather than being in decline as Trump has argued. According to a 2016 PBS *Newshour* report,

> Despite [Trump's] charge that "we don't make anything anymore," manufacturing is still flourishing in America. Problem is, factories don't need as many people as they used to because machines now do so much of the work. America has lost more than 7 million factory jobs since manufacturing employment peaked in 1979. Yet American factory production . . . more than doubled over the same span to $1.91 trillion [in 2015], according to the Commerce Department. . . . Trump and other critics are right that trade has claimed some American factory jobs, especially after China joined the World Trade Organization in 2001 and gained easier access to the U.S. market. . . . But research shows that the automation of U.S. factories is a much bigger factor than foreign trade in the loss of factory jobs. . . . The vast majority of the lost jobs—88 percent—were *taken by robots and other homegrown factors* that *reduce factories' need for human labor.*[58]

The PBS report concludes that the upside of the robot revolution is that the Asian drain of jobs is ending, as manufacturing returns to the US where the use of robots can be easily implemented as an affordable solution to outsourcing. In the technoliberal imaginary, US geopolitical dominance as an economic powerhouse has been reinstated through technological innovation.

The technoliberal argument against Trump's rhetoric of white supremacy both affirms capitalism as the source of US abundance and posits a post-labor economy that heralds a postracial world in which all Americans are equally affected by automation. This move not only erases labor as a useful category of analysis, but in doing so makes it impossible to assess the differential conditions of ongoing worker exploitation under racial capitalism. In this imaginary, technology then stands in for the racialized and gendered devalued work that the colonized, the enslaved, immigrants, and others have performed and continue to perform in the tech economy. As Ruha Benjamin has argued, "if postracial innovators are busily refurbishing racism to make inequality irresistible and unrecognizable, then those who seek radical transformation in the other direction, towards freedom and justice, must continuously reexamine the default settings, rather than the routine breakdowns, of social life."[59] As the preeminent political form asserting postraciality in the present, technoliberalism innovates the fiction of the US as still potentially the site of antiracist struggle against global inequity even as this very formulation updates the ways in which inequity is made manifest economically and socially.

Conclusion: Robo-Trump

Technoliberalism valorizes capital as part of the US postracial condition by negating human labor as a useful category for understanding present-day automation. Yet the expanded exploitation of racialized labor is as central to contemporary automation as it was, for instance, in the context of agribusiness in the early twentieth century discussed by Marez. Human labor, and its differential exploitation under the conditions of racial capitalism, is neither separate from the current tech boom nor a vestige of the past that is now finally overcome with the automation of devalued or degraded work. The prior associations of totalitarianism and fascism with automation in US political and cultural discourses, and the attendant equating of the human-machine and the masses with the politics and aesthetics of Nazism and communism during the Cold War, scaffolds the framing of

Figure 1.5. *M.A.M.O.N.*: Latinos vs. Donald Trump.

technoliberalism as an entirely different moment in technological modernity. The technoliberal imaginary resolves the racial contradiction at the heart of liberalism (namely its entanglement with white supremacy and racial capitalism) by asserting that the postlabor world is also a postrace world. Technoliberalism thus extends the excision of racial violence from liberalism's moral and economic grammars. In light of this update to racial capitalism, we might ask, what would it mean to reassert the centrality of fascism to the technoliberal logics of the present? How might we see technoliberalism as a part of, and not apart from, Trump's America?

Shortly after the November 2016 election of Trump, the Uruguayan creative studio Aparato released a film, directed by Ale Damiani, entitled *M.A.M.O.N.* (Monitor against Mexicans over Nationwide), which portrays Trump as a giant robot at the promised border wall (figure 1.5). This dystopian science fiction short opens with scenes of Mexican Americans performing various jobs, ranging from meteorologist to heart surgeon. Each is shown to be teleported and violently dumped onto a barren desert, on the "other side" of the wall (presumably in Mexico). One by one the heart surgeon, beauty queen, line cook, and schoolchildren crash-land onto the cracked, dusty earth, looking around, clearly confused as to the fate that has befallen them. The line cook angrily approaches the wall first, and exclaims, with passion, "Now what, you fucking gringo? I have my green card!" He approaches what seems to be an outdated intercom system and pushes a button. A robotic voice cheerfully greets him: "US Border Protection. How can I help you?" The line cook replies, "Well yeah, you can help me! Give me my job back." A taco stand crashes on top of the cook's car, and a receipt is slowly printed out of the intercom box. The robotic voice states: "Thanks for your cooperation! And remember, don't come back again." The camera zooms in on the receipt, which turns out to be an invoice charging the expelled cook for the construction of the wall. As the line cook laughs and turns back to the others in the desert to tell them that the "fucking gringo" wants to charge them for the wall, chaos ensues. The wall starts coming down, and a giant, retro robot on wheels made in the image of Donald Trump begins rolling toward the migrants from the US side of the border. As the cook starts to ask Trump for permission to return to the US, a giant steel robotic leg comes crashing down on him, smashing him into the ground. The camera then reveals to the viewers what is inside the robot. Viewed from behind, we see that the robot is being controlled by a man. At the helm of this vehicle is Trump himself, or so the audience can assume based on the signature toupee-like blondish white hair. The man is laughing maniacally. The soundtrack switches to a recording of Trump the candidate promising would-be US voters that Mexico would pay for his "beautiful wall." The speech blares as Trump places a cassette tape (another sign of how retro the robot is) into the cassette deck at the robot vehicle's helm. Missiles then start shooting from the robot onto the defenseless ousted Mexicans. One by one, the expelled migrants attempt to stop Robo-Trump, and dark comedy ensues.

M.A.M.O.N. quickly went viral, garnering 4 million views in less than a few weeks' time. It can be read as a critique of both the white supremacist and technoliberal imaginaries of automation and labor. First, the film insists that racialized labor is central to all aspects of the US economy, from the service industry to entertainment. Thus, the film contradicts the technoliberal fantasy that we are a postlabor society. For example, when the heart surgeon, who is named Dr. Gonzales, is thrown into the desert, he is shown to have left behind one patient, in the middle of a transplant, without a heart. Second, and more interestingly for our argument in this chapter, the film can be read as making a connection between fascist and technoliberal futures.

According to Damiani, depicting Trump as a robot was intended to create "an immense allegorical chariot, a giant stupidity made of sheet metal" full of Asian connotations (a label on the machine reads: "Made in China"). "It is the most ridiculous way we could come up with to spin out the political speech of Trump and give it tangibility," he said. "At the same time it is a retrofuturistic robot, antique and with obsolete technology."[60]

The retro-futurism and technological obsolescence in the film can be read both as a critique of the seemingly outdated nature of Trump's racial and xenophobic rhetoric and as a critique of the logics of newness structuring the technoliberal postracial imaginary. The fantasy of an automated future without devalued, racialized workers is, the film suggests, a fantasy of white nostalgia for a pure US that never really existed (and could not exist) without enslaved, indentured, and immigrant work that is the basis of capitalist development. Moreover, the fantasy of automation is always incomplete. The massive Robo-Trump is not an autonomous machine—there is a man inside the robot (much like the historical Mechanical Turk chess player that actually had a person inside the automaton making all the moves). That the body of Trump and the body of the robot act as one further underscores the dual violence of automation and white supremacy. The film thus refuses to treat them as separate, as does technoliberalism, which proposes technological futures as the solution to Trump's racism and xenophobia. The film, instead, suggests that technoliberal imaginaries that utilize robots to frame antiracist arguments against the logic of walls are violent in their own way.

As we have argued in this chapter, a postrace world forecasted by tech-industry luminaries asserts a universal humanity, thus rendering racialized labor invisible. Because, as we have pointed out, technoliberals cite automation, rather than immigration, as the main threat to the US worker, this occurs ironically through opposition to policies of exclusion, expulsion, and overtly racist discourse. It is thus urgent to pay attention to the connections between the postracial and postlabor logics of liberalism today even as we appear to be in an era ruled by the logics of fascism—the two, we suggest, are never disconnected.

2. Sharing, Collaboration, and the Commons
in the Fourth Industrial Revolution

The Appropriative Techniques of Technoliberal Capitalism

Public Library: Memory of the World is an online archive of digitized books curated by countless anonymous "librarians." The project is collectively organized by hacktivists including Marcell Mars and scholar activists including historian Tomislav Medek. Librarians are contributors who generate free and publicly available digital files of books through their labor of creating, compiling, and cataloguing electronic files of scanned books deemed important for public access by each anonymous individual. Mars describes the process of selection as based on the interest and political commitments of people who volunteer to contribute their labor to the project, rather than on any overarching idea of what texts are most important for public knowledge.[1] Organizers offer workshops and online guides for how to become a librarian and enter books into the online archive through digital scanning, and to aid this process, they have established scanner access in several European cities.[2]

The vision of *Public Library* is that the world's documentary heritage belongs to all and should be preserved and protected for everyone's access. In the menu banner on its home page, the project runs a live log of cata-

Figure 2.1. "Our Beloved Bookscanner."

logued books, which totaled 109,734 on October 31, 2017.[3] Its catalogue is organized by country/language,[4] and chronologically by date of upload.[5] One of the mottos of the project is, "When everyone is librarian, library is everywhere."

Public Library creates librarians and its digital library by installing scanners for digitizing books in publicly accessible locations. It has set up scanning stations like the one pictured here (figure 2.1) in Ljubljana, Zagreb, Belgrade, Berlin, London, and Calafou. The website overview points out that most personal and corporate copiers are also scanners, and with the scanning and uploading software and contribution guides provided on the website (found under headers "workflow" and "tools"), the website promises that "you can learn to be a librarian too. A good one."

In one of the rotating concept notes that run as a banner on the project's home page, the public library is politicized as a historical goal connected to emancipatory revolution, empowering the oppressed by providing the means for them to "reach their dreams." Citing US librarian and educator Melvin Dewey's dream of free schools and public libraries for all, this concept note by Tomislav Medak describes the collective behind *Public Library* as artists and hacktivists and "potentially everyone else" who can make this dream a reality. Of course, discourses of collecting resources for the good of universal humanity have been heavily critiqued by feminists and by postcolonial and critical race scholars, particularly in the arenas demanding repatriations of European colonial museum acquisitions[6] and native remains.[7] Though the project does not align itself publicly with freedom of information movements, it is also compatible with those politics, and therefore is in conversation with critiques of the fetishization of the digital. These concerns must be part of any discussion of common wealth and common good.

The *Public Library* project offers an imaginary of collaboration and sharing that evokes, yet does not adhere to, older socialist imaginaries of the commons and communal good. It provides an example of the proliferation of discourses, hopes, and fears tied to emergent technologies and technological platforms that call for an engagement with how, as Langdon Winner puts it, technologies are "ways of building the world."[8] *Public Library* takes up a long tradition in Western thought of imagining the commonwealth as the antithesis of private property, since at least the early writings of Karl Marx and Friedrich Engels, and therefore as the stronghold of emancipation from capitalist exploitation and appropriation of resources and living labor.

We begin this chapter with the *Public Library* project as a way of introducing how collaboration and sharing through technological platforms can be positioned as part of a progressive politics. Collaboration and sharing, in this sense, are politicized. Meanwhile, these very concepts (collaboration, sharing, and the commons) have undergone a radical decontextualization as they have risen to prominence as descriptors of what makes the technological innovations of the fourth industrial revolution socially and economically revolutionary. Technoliberalism appropriates collaboration, sharing, and the commons to announce capital's unprecedented liberatory potential, while divesting the concepts it uses from an anticapitalist politics and ethos.

In 2015, the *Oxford English Dictionary* introduced "sharing economy" as a term it would now include.[9] The sharing economy is a socioeconomic system built around the sharing of human, physical, and intellectual resources, especially those that individuals may see themselves as possessing and underutilizing (Airbnb, where people rent out unused rooms to travelers, is one well-known example). The sharing economy thus includes collaborative creation, production, distribution, trade, and consumption of goods and services by different people and organizations.[10] The sharing economy is framed as being built on "distributed power and trust within communities [of users] as opposed to centralized institutions," blurring the lines between producer and consumer.[11] Based on the name alone, the much-touted sharing economy, enabled by digital connectivity and wide distribution of the means of production, sounds like it approaches a socialist ideal of "the commons," land or shared resources belonging to a whole community that provide life necessities. Yet, although the sharing economy is sometimes also referred to as the "collaborative economy" because of initiatives based on horizontal networks and the participation of a community, "community" is defined tautologically as simply the whole of those who participate as users. The field of robotics has also taken up re-branding itself. The emergence on the market of collaborative robots has been touted as revolutionizing older, industrial robotics.

In this chapter, we critique technoliberal imaginaries of the so-called creative disruptions to capitalism, which propose that technology will bring about the end of capitalism as we know it through the creation of collaborative robots and a collaborative commons built on the inter-networking of things and people in the sharing economy. We contend that unlike Marxist feminists, who have theorized the rise of capitalist wealth accumulation as dependent on the unpaid labor of racialized and gendered populations, technoliberal appropriation of collaboration, sharing, and the commons reproduces the erasure of racialized and gendered work in their postcapitalist techno-utopias. Within technoliberalism, the commons, once the staging ground and goal of potential socialist proletarian revolution, is evacuated of political content. Sharing becomes an anonymized market transaction that can sidestep the social and what Marx called "species life," a material and ontological underpinning to the commons that gave it a teleologically revolutionary potential. Put otherwise, our critique of the "sharing" in the sharing economy, as our critique of the "collaborative" in collaborative robotics, draws attention to the ways in which the architecture of postindustrial surrogate humanity works through the elision of the racial and gendered dimensions of capitalist development in its production of the fully human. This chapter thus investigates the ways in which socialist concepts of collaboration, sharing, and the commons have been appropriated within technoliberalism for purposes of accumulation and expropriation, even as technoliberalism claims freedom and equality as its express end goal. In contradistinction to the recently popularized discourses of the sharing economy, and to a lesser extent collaborative robotics, imaginaries of technology, sharing, and collaboration from Marx and Engels's *Communist Manifesto* to Haraway's *A Cyborg Manifesto* offer their political motivations as being at the heart of their uses of technology as "decolonizing" and "liberating." At the same time, as feminists and critical race thinkers have pointed out, even these imaginaries risk refusing difference in the name of a universal (humanist) revolutionary call.

The chapter begins with an overview and analysis of how economists, entrepreneurs, and the media have framed the Internet of Things and the sharing economy as a new commons that can liberate human beings from having to perform work considered miserable or degrading. We then connect this context to the emergence of a new category of industrial technologies called "collaborative robots," extending our critique of the imaginary of collaboration in robotics from our critique of how sharing and the commons have been materialized by proponents of the sharing economy. The chapter then moves to a discussion of how racialized and gendered discourses concerning outsourcing in the 1980s and 1990s connect to how categories of race and gender are embedded and embodied (even in their erasures) in the emergence of smart, collaborative robots and the sharing economy. We juxtapose our analysis of how technoliberalism's appropriative techniques uphold the surrogate effect of technology with a dissenting imaginary of the relationship between technology and the commons, and we assess how a project like GynePunk's 3D-printed speculum raises the question of what a collaborative technology that continues a tradition of emancipatory politics and that increases access to knowledge and resources against privatization and property might look like. Rethinking scholarship on technology, liberation, and the commons from Marx and Engels through Haraway, we think through the GynePunk project to reengage a multiplicity of socialist imaginaries against the seeming monolith of technoliberalism's appropriative techniques. We return, as we began, to *Public Library*, in order to dwell with alternative imaginaries of use that don't have to do with the propagation of capitalist production, and what radical collaborations and collaborative commons are possible outside the technoliberal imaginary.

A 2015 report released by Bank of America claimed that the three main "ecosystems" of "creative disruption" in the global economy are the Internet of Things, in which ordinary objects such as thermostats have network connectivity that enables them to send and receive data; the sharing economy; and online services.[12] These ecosystems, along with what Bank of America termed a revolution in robotics—the prediction that by 2025, 45 percent of all manufacturing jobs will be done by robots—will, according to the report, profoundly change "how we work and live."[13] A second report issued by the Bank of America further estimates that

> The robots and AI solutions market will grow to US$153bn by 2020, comprising US$83bn for robot and robotics, and US$70bn for AI-based analytics. . . . Adoption of robots and AI could boost productivity by 30% in many industries, while cutting manufacturing labour costs by 18–33%. We are facing a paradigm shift which will change the way we live and work. We anticipate the greatest potential challenges, . . . notably the possible displacement of human labour (with 47% of US jobs having the potential to be automated) and growth in inequality (with a 10% supply and demand gap between skilled and non-skilled workers by 2020).[14]

The "robot revolution" and the sharing economy, based on the internetworking of objects and people that this Bank of America report highlights, are both a part of what the World Economic Forum has termed the Fourth Industrial Revolution and what scholars Erik Brynjolfsson and Andrew McAfee have termed "the Second Machine Age."

According to Klaus Schwab, founder of the World Economic Forum, "The First Industrial Revolution used water and steam power to mechanize production. The Second used electric power to create mass production. The Third used electronics and information technology to automate production. Now a Fourth Industrial Revolution is building on the Third, the digital revolution that has been occurring since the middle of the last century. It is characterized by a fusion of technologies that is blurring the lines between the physical, digital, and biological spheres."[15] Similarly, for Brynjolfsson and McAfee, the industrial revolution can be thought of as the first machine age, while the second machine age, driven by Moore's law (Gordon Moore's 1965 prediction that "computing would dramatically increase in power, and decrease in relative cost, at an exponential pace"), is about the automation of cognition evident in innovations such as Google's autonomous cars and IBM's Watson computer, which can beat the best human *Jeopardy!* players.[16] Unlike the World Economic Forum, which is hopeful about the human benefits of the fourth industrial revolution, however, Brynjolfsson and McAfee, along with the Bank of America report, caution that as machines replace humans in various spheres of life and labor, there will be increased inequality—something that is already visible in the economies of the US and Western Europe. At the same time, they remain hopeful about programs that could retrain humans toward different potential roles.

The coming together of techno-utopic and more cautious approaches to the economic paradigm shift brought about by technological infrastructures is elucidated in the compendium of short essays on technology and the future of work put together by *Pacific Standard* in 2015. This project assembled speculations on the topic from industry leaders, technologists, social scientists, and journalists. One focus in these debates was the contest between technologies designed to enhance what humans already do and technologies designed to replace humans. In spite of differences, both sides tether techno-futurity to the surrogate human effect of emergent technologies and digital platforms. John Markoff (science writer for the *New York Times*) describes two primary arguments about the effect of technology on labor: (1) AI and robotics are developing so rapidly that virtually all human labor will be replaceable within three decades (represented for him by Moshe Varde at Rice University); and (2) robotics will lead to new jobs and create immense economic activity (represented by the International Federation of Robotics). In both perspectives, human futures are linked to capitalist-driven technological developments focused on questions of productivity and efficiency.

The two philosophies of how to engineer human–robot and human–AI relations in the realm of labor presented by Markoff date back to 1964 and the beginning of interactive computing at Stanford.[17] John McCarthy, who coined the phrase "artificial intelligence," believed he could begin designing an AI that could replicate human capabilities in a single decade. In contrast, Douglas Engelbert set up a lab with the goal of developing technologies that would "augment," or extend, rather than replace, human capabilities.[18] These evolved into two opposing camps that worked largely in isolation: artificial intelligence research and human–computer interaction design research. Boston Dynamics, the MIT Media Lab–derived US military robotics contractor, is the preeminent example of the AI group. Tom Gruber, an Apple engineer who designed Siri speech recognition, has worked to prototype the other. The competition between two models of a future dominated by machines that Markoff sets up brings attention to the way they will materialize existing implicit and explicit social values: "Will a world watched over by what the 1960s poet Richard Brautigan described as 'machines of loving grace' be a free world? The best way to answer questions about the shape of a world full of smart machines is by understanding the values of those who are actually building these systems."[19] In short, engineers and designers will build values into the hardware and software constituting the infrastructure that organizes human–machine interaction.

Because they are sutured to capitalist demands for faster production to generate more profit, present-day techno-futurities, while claiming to be about technologies that are distinct from the modes of automation in the early and mid-twentieth century we addressed in chapter 1, continue to be haunted by the specter of the obsolescence of the human worker. Thus, a number of scholars and thinkers still fear that this second machine age will drastically reduce the number of available jobs, leading to more and more people without employment. Computer scientist and leader in artificial intelligence development Nils J. Nillson sums up these concerns in his assertion that automation will put wealth in the hands of a small number of "super-managers" worldwide, while leaving the masses jobless and impoverished.[20] The question of what will happen to the vast numbers of unemployed people seems to be up for all sorts of dystopic conjecture: working more and more for less and less, perhaps winning more leisure time (but will people start hobbies or will they just do drugs and other socially destructive things? asks Nillson). While Nillson positions techno-dispossession as a thing of the future, it is not difficult to find real-time models for economies that cannot employ their own citizens. Most of the formerly colonized world contends with this problem in different ways because of international lending (Third World debt) and the ongoing struggle to regain self-sufficiency in restructuring infrastructures designed to evacuate resources. A small number of people will gain wealth, as in India and China, but the vast majority will lose quality of life, lose land, and lose employment. There will be massive migration to centers of wealth, and modes of life and self-sustenance will be further destroyed, then rebuilt and managed for the gain of others, elsewhere.

For example, in the issue of *Pacific Standard* on this topic, contributors suggest that one solution to the coming robotics-induced obsolescence of human workers is to institute a universal basic income. As robots take our jobs, the argument goes, the additional wealth created will be enough that the growing number of unnecessary human workers can be supported. Even some longtime labor activists, most prominently Andy Stern, the former president of the largest and one of the most influential unions in the US, the Service Employees International Union (SEIU), which represents 1.5 million public service workers, including nurses, care providers, and security workers, have capitulated to the idea that within the next two decades machines will replace over half of US jobs.[21] According to Stern, as more tasks are automated and full-time jobs disappear, the role of collective bargaining will become marginalized and dues-paying union members will be fewer and farther between. Stern is a member of the steering committee of one of the foremost organizations pushing for a universal basic income (UBI), the Economic Security Project, which asserts that "In a time of immense wealth, no one should live in poverty, nor should the middle class be consigned to a future of permanent stagnation or anxiety. Automation, globalization, and financialization are changing the nature of work, and these shifts require us to rethink how to create economic opportunity for all."[22] To enable a new American Dream, argues Stern, we need to implement a UBI of at least $12,000 a year so that no one falls into poverty. However, as union educator, activist, and scholar Heidi Hoechst points out, when all that remains of a labor movement is the fight for a UBI, the movement has capitulated to the neoliberal restructuring of the fabric of society.[23] With everyone receiving a minimum income, the last vestiges of social support, such as the costs of Medicare and welfare, will be transferred to the individual.

What will happen to workers with the increase of robotic automation in manufacturing is a question that news media and technoliberal elites are also rushing to answer.[24] Christopher Hughes, the cofounder of Facebook, is also the cofounder of the Economic Security Project. He argues that in a time when the US is more divided than at any other time since the Civil War, and in a time when faith in the opportunity for a good life in the US has waned, it is crucial to use cash in the form of a distributed universal basic income to harness technology for social justice. As he puts it, "Median household incomes in the US haven't budged in decades even though the price of healthcare, education, and housing have skyrocketed. The old idea that if you work hard and play by the rules, you can get ahead has disappeared. As a handful of people at the top have thrived, the rest of America—urban and rural, white people and people of color, old and young—has nearly uniformly been left behind."[25] While Hughes acknowledges that historically the myth of opportunity has not applied to people of color in the US, in his formulation of an argument for UBI, precarity has become universal: "Americans of nearly all backgrounds now believe their kids are likely to fare worse than they have. Major forces like automation and globalization have changed the nature of jobs, making nearly half of them piecemeal, part-time, and contingent—Uber drivers, part-time workers, TaskRabbit workers abound."[26] In the technoliberal formulation of UBI, which would allow everyone to, in Hughes's words, "create a ladder of economic opportunity," the regressiveness of UBI is justified because what used to be racialized precarity now affects all (including white people who were formerly securely in the middle class because of race privilege). It is thus an appropriation of both the racial histories of devalued labor and a socialist imaginary of distributed wealth. In fact, however, the most that is proposed as part of UBI is $1,000 per month—hardly a subsistence wage in most parts of the United States.

Universal basic income is particularly interesting in thinking about how and why socialist ideals are redefined and appropriated as part of technoliberal reimaginings of the common social good. Hughes and others are aware that the increasing wealth disparity and the disproportionate number of young tech billionaires whose wealth accumulation has been unprecedented are unsustainable and unjust. Yet, in articulating a need for UBI, they also assert that technology is a kind of public good tied to US citizenship. In a 2017 article in *Wired* magazine, two examples are used to explain why UBI can work. The first is a survey of Alaskans who receive $2,000 per person per year from the state's oil revenues conducted by the Economic Security Project. In this example, the profit from oil, a public good, is distributed among residents of the state (thus membership in the state determines equal distribution). The second example is that of the Eastern Band of Cherokee Indians in North Carolina, who split 97 percent of casino profits. The model is slightly different from that of Alaska: "In 2016, every tribal member received roughly $12,000. . . . All children in the community, have been accruing payments since the day they were born. The tribe sets the money aside and invests it, so the children cash out a substantial nest egg when they are 18."[27] At the moment, this payout is around $105,000. The article, rather than dwelling on the politics of land, dispossession, settler colonialism, and the politics of tribal casinos in the US, instead asserts that citizenship, rather than need, should be the basis for distributing UBI:

> The idea is not exactly new—Thomas Paine proposed a form of basic income back in 1797—but in this country, aside from Social Security and Medicare, most government payouts are based on individual need rather than simply citizenship. Lately, however, tech leaders, including Facebook founders Mark Zuckerberg and Chris Hughes, Tesla's Elon Musk, and Y Combinator president Sam Altman, have begun pushing the concept as a potential solution to the economic anxiety brought on by automation and globalization—anxiety the tech industry has played its own role in creating.[28]

In the technoliberal imaginary of a just distribution of some of the tech wealth, then, settler citizenship both appropriates and erases the settler colonial violence upon which wealth accumulation is based in the US. Notions of distribution (and of tech as a US national resource) thus also further the project of US imperialism.

Colonizing the Commons

In spite of claims that we are entering a new social and economic moment that, enabled by technology, can finally fully free the human from drudgery of service work and repetitive tasks that can now be done by machines, the reliance on surrogates who perform devalued tasks to enable the freedom and autonomy of the liberal subject is one that dates back to imperial modernity and the very premises of capitalist developmentalism. The ways in which the socioeconomic scaffolding of empire enabled imaginaries of surplus or obsolete humanity to begin with is crucial for understanding the dystopic fears surrounding the surplusing of modern man himself.[29] We might think of this as the longer historical context for Stephen Hawking's publicly expressed fear that AI will be the end of the human species.[30] At the same time, techno-utopic predictions about the end of capitalism, the focus of this section, as well as less optimistic assessments of the second machine age that fear a rising tide of unemployment as discussed above, displace humanity from the scene of degraded, racialized, and gendered work (enabled through technological development) in order to posit a moment of emancipation or species evolution.

Like universal basic income, the so-called sharing economy is an example of technoliberalism's postlabor imagination of common wealth, or a common social good, that is in fact a technoliberal appropriation of socialist ideals that further the capitalist expropriation of labor and life. For example, Jeremy Rifkin's popular book *The Zero Marginal Cost Society* argues that the internetworking of humans and things—physically and affectively— renders human labor obsolete but will in fact ultimately revolutionize life by freeing humans for more meaningful or creative pursuits. Rifkin is a US economic and social theorist who has written over twenty books since the 1970s on the environment, energy, the economy, and technology, and who has advised numerous world leaders in Europe. He is "the principal architect of the European Union's Third Industrial Revolution long term economic sustainability plan to address the triple challenge of the global economic crisis, energy security, and climate change."[31] It is from this stance that he builds on his prior interest in political-economic alternatives to posit that technological innovation can bring about a zero marginal cost society and an end to capitalism. This is Rifkin's formulation of a new collaborative commons—one based on human creativity and collaboration enabled by technological advancements and the infrastructures of the sharing economy.

The Internet of Things (IoT) is a social-technological infrastructure that is designed to largely bypass the need for human oversight and intervention, and yet manage the mundane and reproductive work of daily life. This seemingly neutral and mechanical technological infrastructure, composed of so-called smart objects that communicate with one another, organizes the temporal experience of work and the form of subjectivity through which one must engage that infrastructure. The IoT (see figure 2.2) has been touted by engineers and writers as the next economic paradigm shift, and Rifkin has been its biggest proponent, heralding it as bringing about human emancipation from work and the end of capitalism. In *The Zero Marginal Cost Society*, he writes, "If I had told you 25 years ago that, in a quarter century's time, one-third of the human race would be communicating with one another in huge global networks of hundreds of millions of people . . . and that the cost of doing so would be nearly free, you would

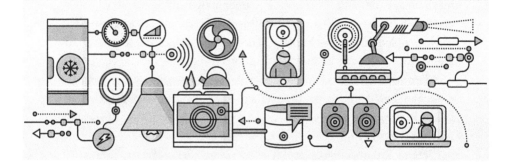

Figure 2.2. The Internet of Things as depicted by *Computer Weekly*.

have shaken your head in disbelief."[32] Rifkin takes the notion that free information and communication are harbingers of a large-scale revolution in which we move toward a "near zero marginal cost" society—one in which "nearly free goods and services" emerge through the optimization of productivity (that is, with the development of technologies such as 3D printing, because of which the "cost of producing an additional good or service is nearly zero").[33] In this postcapitalist techno-utopia, with goods being free to produce, the very idea of property would become meaningless. Meanwhile, traditional areas of work, even that of repairing the machines that create goods and render services, would be done by the machines themselves.

Rifkin sees smart technologies as enabling the uncoupling of human productivity from employment, thus freeing humans for the "evolving social economy" embedded in a "Collaborative Commons" organized by social networks and open-source programming. "Big Data, advanced analytics, algorithms, Artificial Intelligence (AI), and robotics are replacing human labor across the manufacturing industries, service industries, and knowledge-and-entertainment sectors, leading to the very real prospect of liberating hundreds of millions of people from work in the market economy."[34] Rifkin argues, in short, that the infrastructure revolution, marking a break from the first (eighteenth-century) and second (early-twentieth-century) industrial revolutions, emancipates human creativity from the drudgery of wage work.[35]

Although this argument is ostensibly about technology as the condition of possibility for freeing human creativity, Rifkin glosses over the question of how economic, social, and human obsolescence has been figured through a racial-imperial episteme. The replaceability of human labor through the surrogate effect of objects in the Internet of Things is contextualized by capitalist development in the Global North, in which the specter of unemployment is attached only to those populations not already marked for elimination or surplus. The Internet of Things thus effectively materializes assumptions of what constitutes a human, even as it excludes those who are not the intended subjects of a postlabor world. The celebration of the Internet of Things and "smart" infrastructures is connected to colonial spatial imaginaries that are the foundation of the sharing economy. We identify these technologies as part of an impetus to colonize the notion of the common in the name of proclaiming the end of capitalism. New technologies that purport to substitute smart objects for humans extend racial colonial logics into the imagined evacuation of the human entirely. As discussed in the introduction to this book, the condition of surplus being, or of the production of the obsolete, is always racialized even if, and precisely because, techno-revolutionary fantasies of the twenty-first century congeal around a dehistoricized imaginary of emancipation and free labor.[36]

The gender component of this structuring contradiction—whereby the populations subjugated through racial and gendered differentiation are surplused, and their protest and struggle to preserve the commons or resources of common wealth are hidden in favor of a teleological narrative of capitalism's historical development—is at the center of Silvia Federici's *Caliban and the Witch*. This study, in the tradition of materialist, socialist, and Marxist feminisms, points to this structuring contradiction in its assessment of the end of the commons and the ascent of global capitalism. Federici argues that the privatization of common lands (the enclosure of the commons) necessitated the taking away of women's autonomy over their bodies and the use of gender and class as a means to divide and conquer opposition to primitive accumulation.[37] She also connects the emergence of reproductive unpaid women's work in Europe, which enabled the accumulation of wealth, to European expansion in the colonies. These substitute (or, we might say, surrogate) workforces were meant to replace the rebellious peasants opposing the enclosure of common lands and resources. Demonstrating that the commonplace assumption that modern capitalism defeated the tyranny of feudalism is a mythology that persists today, Federici argues instead that what was in fact defeated by capitalism and privatization were mass uprisings and opposition to the enclosure of the commons. The sixteenth- and seventeenth-century witch hunts in Europe, she suggests, can therefore be read as attempts to quell what was the first women's movement for bodily autonomy.

Federici's argument is crucial for understanding how certain figures (like the witch) must be expunged from the history of modern capitalism so that, rather than being seen as figures of rebellion, they become the spectral figures of an unenlightened premodernity. As we argue here, the new collaborative commons, rather than being based on earlier socialist models of the commons, is instead built upon these very erasures that enabled the spread of global capitalism. This ethos of capitalist expansionism, enabled as it is by the division and capture of racialized and gendered labor, persists in the present-day techno-utopic vision of a new commons. We see this, for instance, in the way Rifkin distinguishes two categories or species of the human produced within the IoT: There are the makers, and there are those who are replaceable by self-replicating machines, such as 3D printers.

With regard to the former, Rifkin contends that the maker movement is nothing less than the early embodiment of the new human consciousness. He cites the creation of Fab Lab in 2005 at MIT, an outreach project that uses open-source software to allow anyone to create their own 3D printed objects, thus creating prosumers (producers and consumers).[38] The now seventy Fab Labs, though primarily located in the industrialized urban North, have also made their way to the Global South, where they enable simple tools and objects to be created in order to advance economic welfare.[39] Purportedly emanating from "world class universities and global companies," Rifkin calls this multisited space the "people's lab" as it has made its way to nonelite neighborhoods. In this frame, 3D printing is understood to be part of the new collaborative commons because of its potential to democratize the means of production. Equally important, Rifkin highlights that the cost of making thousands of products is no more than the cost of making just one. A dream machine, the 3D printer can even make its own parts, rendering it a "self-replicating" machine that costs next to nothing to repair and run.[40]

Rifkin envisions more than just a trickle-down benefit of creativity from the urban Global North to the Global South, which can then use these innovations of the collaborative commons for economic uplift. He also understands technological self-replication and cheapness to lead to a reversal of outsourcing, which has been the hallmark of imperial and neoliberal economic practice, and which relies on surplus populations that it has itself made surplus. Yet, as his thinking demonstrates, these populations must be rendered obsolete in order for the IoT economy to thrive. Rifkin writes, "manufacturers that have long relied on cheap labor in their Chinese production facilities are bringing production back home with advanced robotics that is cheaper and more efficient than their Chinese workforces. At Philips's new electronic factory in the Netherlands, the 128 robot arms work at such a quick pace that they have to be put behind glass cases so that the handful of supervisors aren't injured. Philips's roboticized Dutch electronics factory produces the equivalent output of its Chinese production facility with one-tenth of the number of workers."[41]

That one species of human (the exploitable labor force in former Second and Third Worlds) is being replaced in its function by technological innovation—by things (surrogate human technologies) in the IoT—replicates a sliding scale of humanity established through the development of capitalism in the colonies. In this sense, the use of robots as replacements for degraded workers confirms an already existing bias about what kind of workers can be replaced easily by machines that are more accurate and economical. Terry Gou, CEO of Foxconn, articulates the racialized condition of replaceability most explicitly. Gou, Rifkin relates, "joked that he would prefer one million robots [to his one million workers]. 'As human beings are also animals, to manage one million animals gives me a headache.'" As Rifkin elaborates, "China, India, Mexico, and other emerging nations are learning quickly that the cheapest workers in the world are not as cheap, efficient, and productive as the information technology, robotics, and artificial intelligence that replace them."[42]

The "new human consciousness" of the collaborative commons that Rifkin predicts will emerge through the makers movement thus reasserts a human–thing hierarchy that is mapped onto well-worn racial–imperial paradigms.[43] Yet, despite techno-utopic projections, as we elaborate in chapter 3, human labor power continues to be an irreplaceable commodity, highlighting the growing unevenness between racialized, gendered, and ostensibly endlessly exploitable populations who labor in places like China, India, and Mexico. While it is worthwhile to point out how neatly Rifkin's techno-utopia relies on imperial and neoimperial imaginaries of surplus labor (as surplus populations), it is equally important to note the need to replace the exact functions of such populations with technology. This is a racial logic of elimination in which the violence happens not through physical extermination. Rather, the violence of this fantasy occurs through the desire to subsume the global racial other into the IoT's "things" by reducing the "cheapest" of labor subjects to their mere function within global capitalism.

Collaboration and the Robot Revolution

The Internet of Things and other techno-fantasies may be entangled with wishes to be freed from miserable labor, but even when they have a political intention like the end of capitalism, they reproduce a worldview in which technology is conceived and developed only to do what humans already do. Put otherwise, technologies that scaffold the fourth industrial revolution and second machine age are reproducing the ongoing contradictions and violence of capitalist expansion, imagining and generating only surrogate humans, rather than new social, political, and subjective forms yet to be. One recent development in robotics, to which we turn in this section, that attempts to resolve the anxieties about robots replacing humans (and of racialized outsourcing of cheap labor) is the emergent field of collaborative robotics. Collaborative robots are said to work alongside humans, rather than being designed to replace them. Yet "collaboration" as imagined in mainstream robotics rehearses the mind/body split, where the human can now be free to cultivate the soul (that is, creative potential) because it is liberated from unnecessary toil. Meanwhile, the robot does the devalued work of the body. When we examine the history of collaborative robot design, we see an insistence on affirming an appropriate level of difference in the form and function of varieties of labor performed by humans and machines. This is a level of difference that makes machines unthreatening because they maintain the primacy of the human. The problem of degrees of difference between human and machine (that is, the distance along a spectrum from the fully human to the unquestionably nonhuman) raises the critical question of what it is that a human does (and how the human is defined by the quality of their labor).

Maintaining a clear idea of human essence through an articulation of what is different between what humans and machines can and should do is a defining aspect of collaborative robots. As technologies that enact the surrogate human effect, collaborative robots take up the racial imaginary of labor without intelligence that is also at work behind the logics of colonization and outsourcing. Indeed, emergent technologies and platforms propose a future free from degraded work through "robot collaboration," yet the infrastructures of the collaborative human–machine economy retain the degraded categories of labor formerly done by racialized others. Hopes that the surrogate human effect of technology will free "humans" from degraded labor thus also necessitate platforms that must actively conceal the fact that other forms of "miserable" work are still being done by humans. The surrogate human as collaborator extracts miserable work out of populations marked for elimination or extinction (factory workers and the racialized low-wage laborer) even as it substitutes the partner and tool for the surrogate human. As Langdon Winner has written, a flexibility that inheres in technologies vanishes "once initial commitments are made" as to how these technologies will order human activity and influence work, consumption, communication, and everyday life.[44]

Models of a postlabor technological revolution via smart objects and the sharing economy subsume "collaboration" and "sharing" in the name of a neoliberal interest in increased exploitation of "free" time. The collaborative parameters of collaborative robots, much like the sharing aspect of the sharing economy, seem to mark a complete shift from earlier socioeconomic models of labor exploitation, but in reality they demand an ever-increasing use of spare time and unpaid labor to be put back into the economy of things. For instance, in September 2015, the BBC ran an interactive quiz titled "Will a Robot Take Your Job?" Part of a longer series of articles, videos, and interactive content analyzing what the news outlet termed the "revolutionary" nature of AI and robotics in the twenty-first century, occupations ranging from actor to tax expert were classified by percentage of "risk" for human obsolescence. Selecting probation officer from the list, for instance, a user would find that this profession is quite unlikely to be replaced. In contrast, telephone salespeople, bookkeepers, and waiters/waitresses are categorized as almost certainly obsolete professions for humans.

What was the reasoning behind these classifications? According to the BBC, "Oxford University academics Michael Osborne and Carl Frey calculated how susceptible to automation each job is based on nine key skills required to perform it: social perceptiveness, negotiation, persuasion, assisting and caring for others, originality, fine arts, finger dexterity, manual dexterity and the need to work in a cramped work space."[45] The study concluded that "roles requiring employees to think on their feet and come up with creative and original ideas, for example artists, designers or engineers, hold a significant advantage in the face of automation. Additionally, occupations involving tasks that require a high degree of social intelligence and negotiating skills, like managerial positions, are considerably less at risk from machines according to the study." In contrast, the study explains, "while certain sales jobs like telemarketers and bank clerks may involve interactive tasks they do not necessarily need a high degree of social intelligence, leaving them exposed to automation. As more advanced industrial robots gain improved senses and the ability to make more coordinated finger and hand movements to manipulate and assemble objects, they will be able to perform a wider range of increasingly complex manual tasks."[46]

Twenty-first-century reconfigurations of human–machine social relations continue to be haunted by specters of unemployment, echoing the historical crises surrounding technological development and instances of racialized outsourcing of forms of labor. At the same time, as with earlier racial discourses about unemployment, contemporary articulations propose that there is a fully human essence that can never be replicated by the nonhuman or not-quite-human. The BBC quiz, with its distinction between modes of labor that can be replicated by robots and those that cannot, recalls the racial panic brought about by outsourcing that began in the early 1990s. Outsourcing, as a technique of cheapening production and service labor, often moved operations to the same decolonizing labor populations in the Third World and Global South. The justification was that these laborers were most fit for reproducing the inventions of the centers of capital in the US and northwestern Europe.[47]

For MIT roboticist Rodney Brooks and others, the key to allaying techno-dystopic anxieties of impending human obsolescence through robotics and outsourcing is to focus on developing robots that are different from but complementary to the function of the human worker. Brooks's robots are not surrogate humans, but, as the replacement of miserable functions that were never human (the so-called dull, dirty, repetitive work), they enact a surrogate effect undergirding the fantasy of human freedom. As the engineering and functionality of collaborative robots makes clear, seemingly novel notions of human–machine collaboration can still perpetuate the human social relations of racial colonialism as new technologies inhabit a space of subservience. This is evidenced in their programmed responsiveness and receptiveness to others' commands

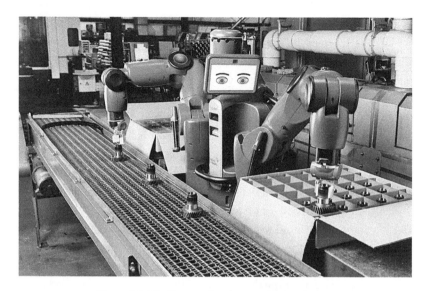
Figure 2.3. The Baxter robot from Rethink Robotics.

and desires as workers. In 2012, Brooks, best known for his research on robotics and mobility, and commercially for his participation as a co-founder of iRobot (maker of the Roomba vacuum cleaner), introduced the world to Baxter (figure 2.3). Baxter, which is a product of Brooks's most recent for-profit venture, the firm Rethink Robotics, is billed as "the safe, flexible, affordable alternative to outsourced labor and fixed automation."[48] Baxter is the best-known collaborative robot on the market to date. "A cobot or 'collaborative robot' is a robot designed to assist human beings as a guide or assistor in a specific task. A regular robot is designed to be programmed to work more or less autonomously."[49] In the present moment, Cold War–era robots have been refigured as dumb and dangerous to workers—large, heavy, difficult to program, and so potentially hazardous to the workforce that they had to operate behind cages. As roboticists like Brooks insist, there is nothing to fear in factory robots 2.0 like Baxter. After all, according to this line of argumentation, none of these earlier epochs of machine evolution displaced humans from the workforce—rather, they just changed the nature of what humans do (on factory floors or in the office).[50]

Baxter has been advertised as redefining the parameters of human labor. Rethinking robotics, as in the company's name, conjures the need to also rethink the nature of human work. Brooks and the media coverage he has received have underscored that Baxter is designed to work alongside humans. This is what makes it a collaborative and complementary machine that ostensibly enhances the human, rather than a machine designed to replace the human. In this sense, Brooks proposes that contemporary robotics break away from Cold War industrial robots discussed in the previous chapter. Collaboration in the field of robotics, like the sharing of the sharing digital economy, refigures notions of the collective by looking for ways to exploit and discipline untapped human potential, now "freed" through technology. In this imaginary, older Marxist and socialist perspectives on labor as a state of political consciousness, and as a condition around which to organize political action, are rendered obsolete, along with the older model of the worker who has been marked for extinction. Collaborative robots like Baxter seek to usher in a relation between robot and machine that enhances the human, now divorced from labor as a model for collective action. Indeed, the collaboration between the robot and the human makes a human workforce unlikely subjects of rebellion against exploitation.

Baxter is a humanoid robot with a tablet serving as its "head." Because of the revolutionary attributes attached to Baxter as a new kind of industrial machine, we describe its form and function in some detail. Baxter's tablet head displays

animated cartoony eyes and eyebrows [that] change expression to convey intuitive messages to nearby workers. These expressions include closed eyes when it is on stand-by; neutrality when it is ready to start learning (eyes wide open and eyebrows parallel to eyes), concentration when its [sic] in the middle of learning (eyes wide open and eyebrows slanted down toward the center), focus when it is working without any issues (eyes looking downward and eyebrows slanted down toward the center), surprise when a person comes nearby (wide eyes with dilated pupils, eyebrows raised and an orange screen background), confusion when it's having an issue with a task (wide eyes with one eyebrow inverted and both slanted downward toward the outside) and sadness when there is a problem and it has given up trying to work on a task (eyes looking downward with both eyes and with eyebrows inverted and slanted down toward the outside). Baxter's eyes also move in the direction one of its arms is about to move as a warning to anyone working nearby.

On its torso, two arms are mounted that measure 41 inches (104 centimeters) from shoulder to end-effector plate, an area with interchangeable end-effectors, or "hands," that serve various purposes. Baxter comes with two hand types: a vacuum cup and an electric parallel gripper with something like fingers. . . . Each arm has seven degrees of freedom, and they can work in tandem or independently of one another, depending upon need. This means you can do things like put Baxter in between two conveyor belts and have it work both. The arms are also compliant, a robotics term meaning that rather than being completely rigid and unable to change course, they can sense and adjust to any obstacles they encounter. If you grab, push or bump into one of its arms, it will give rather than remaining fixed. This compliance is made possible by using series elastic actuators, in which a motor and gearbox control a spring that drives the joints, rather than directly controlling the joints. It's the springs that make each arm less rigid than typical robot arms, and they are also used to measure forces acting on the arms.[51]

According to Brooks, the idea for the Baxter cobot came about through his travels to China, where the iRobot products were manufactured. In an interview, he explained that he "'realized that [outsourcing manufacturing to China] wasn't sustainable, because once the cost of Chinese labor starts to go up, the appeal of doing a product there starts to go away. . . . He concluded that a fairly simple robot could do lots of those tasks, like basic material handling, packing and unpacking boxes, and polishing and grinding."[52] Thus, Baxter, like other collaborative robots designed to work alongside a human workforce, was built to replicate "repetitive and boring jobs, and ergonomically challenging tasks," so that these tasks could be done right in the US and Western Europe.[53]

The human–machine partnership embodied in the collaborative robot, while celebrating the emancipation of the worker in the Global North, nonetheless presumes and necessitates the excision of the global racial other (the surplus is here initially articulated as those distant from Euro-American modernity rather than the worker). Brooks's spark of inspiration during his travels to Chinese factories sheds light on the racial infrastructure of robotics innovations in the field of labor. On the one hand, the impetus toward the invention of Baxter suggests that as a surrogate for outsourced labor, Baxter will return manufacturing to the US. This is something Brooks has articulated as an explicit goal of the machine.[54] At the same time, the racial/devalued imaginary of labor appropriate for outsourcing remains intact. As the Rethink Robotics webpage touts, "This smart, collaborative robot is ready to get to work for your company—doing the monotonous tasks that free up your skilled human labor to be exactly that."[55] This statement suggests that not much has changed from the 1990s distinction between creative, skilled labor and the uncreative, monotonous, reproductive labor that undergirded understandings of how outsourcing would work.

Other kinds of collaborative robots, like Botlr, are also designed to fill in for tasks seen as undesirable for a human workforce (figure 2.4). As one journalist observed after receiving a courtesy toothbrush delivered by the short robot who moves around on wheels and has a touchscreen as its head, "Botlr isn't the first mass market robot, but it is among the first to perform a service industry job that was once exclusively done by humans. Work as a bellhop may not be the most appealing job, nor does it pay particularly well. But usually, it's a person's job. In this Aloft Hotel, at least, a person isn't needed anymore."[56] Yet, as the piece on Botlr concludes, those marginalized skills that can be "robot replaced" need not signal the mass displacement of humans by robots. Rather, Botlr is an indication that human society needs to ask how it can be "robot enabled" rather than "robot replaced."[57]

The work of enabling and improving the lives of privileged subjects, it is worth recalling, is historically and geopolitically racialized and gendered. Those conceived of as already fully human are never threated with *replacement* by a mass of exploitable natural and human resources; rather, those resources *enable* their lives as fully human (and indeed, what Vora has elsewhere termed their vital energies for creativity, productivity, and even reproductivity). It is thus difficult to ignore the global racial imaginaries built into notions of robots that can undergird human wonder at the marvels of technology in places like a Cupertino hotel next door to Apple's headquarters, where Botlr serves guests. In this sense, we might view Botlr as the inheritor of the Bangalore Butler. As Vora writes, in an earlier moment, 2007, the Bangalore Butler was touted as the newest development in the outsourcing of personal assistants. She explains that not only is the Bangalore Butler "redolent with a [racial] fantasy of the luxury of British

Figure 2.4. Botlr at the Aloft Hotel in Cupertino.

colonial India, where brown men in crisp white uniforms and turbans served meals on silver platters to smartly dressed colonials," but also that "Indian workers occupy particular positions in the international division of labor as a result of the material conditions India inherited from the British colonial period."[58]

The kinds of skills and the kinds of people marked as replaceable, therefore, are already entangled with the historical weight of the kinds of skills and people that are rendered valueless precisely because of their exploitability in an imperial economy. Indeed, as even the most prominent proponents of human–robot collaboration acknowledge, some skills and certain people *are* being replaced. For instance, those performing outsourced work are now ostensibly made redundant by robots (though it is more the imaginary that they are rendered redundant, rather than their actual redundancy, that fills out the racial form of our contemporary techno-imperial epoch).

To return to Baxter: like the Chinese workforce that inspired Brooks, Baxter is billed as cheap. The cobot only costs $22,000 for a business to acquire (as opposed to $50,000–$100,000 for conventional industrial robots), and the cost of Baxter's "employment" is only $4 an hour.[59] When the robot becomes even cheaper than an already disposable reserve workforce, it makes clear the extent to which attributes of the properly human are not only racial, but also geopolitical. In the cobot imaginary, properly human labor that has been "freed" by machine labor still resides primarily in the (post)imperial US and the industrialized Global North.

On the Extinction of the Industrial Worker in a Postlabor World

We have argued thus far that collaborative robots in factories refigure the robot as an emancipating tool and partner for the human as the latter evolves (with the help of technology) toward less mundane, boring, dirty, dangerous, and repetitive tasks. In this section, we consider how the notion of human–robot "collaboration" also works to predict the extinction of a particular species of the human: the industrial worker of an earlier era of automation. The specter of a dying workforce as a way to refigure the "scariness" of the robots that will take human jobs and shift toward a more humane version of robots poised to not only collaborate with, but care for, a human population is central to how human–robot collaboration has been framed in the twenty-first century. Much of the publicity around robot–human collaboration centers on robots as already being embraced without reservation by US corporations and workers. As George Helms of Steelcase said of Baxter and its successor robot, Sawyer (a lighter one-armed robot), these are tools meant to be employee "enhancements" and "multipliers" rather than replacements.[60] Significantly, the cobot is not seen as just a tool for the performance of certain tasks; in addition, it is framed as a tool in the evolution of what a worker does. Put otherwise, worker sociality and sense of collectivity are tethered to the robot, and not to fellow workers.

Throughout his interviews and talks about Baxter, Brooks has painted a picture in which factory workers come to embrace collaborative robots. In the following speech, he tells the story of Mildred, an older worker who has grown to love factory robots:

Mildred's a factory worker in Connecticut. She's worked on the line for over 20 years. One hour after she saw her first industrial robot, she had programmed it to do some tasks in the factory. She decided she really liked robots. And it was doing the simple repetitive tasks that she had had to do beforehand. Now she's got the robot doing it. When we first went out to talk to people in factories about how we could get robots to interact with them better, one of the questions we asked them was, "Do you want your children to work in a factory?" The universal answer was "No, I want a better job than that for my children." And as a result of that, Mildred is very typical of today's factory workers in the U.S. They're older, and they're getting older and older. There aren't many young people coming into factory work. And as their tasks become more onerous on them, we need to give them tools that they can collaborate with, so that they can be part of the solution, so that they can continue to work and we can continue to produce in the U.S. And so our vision is that Mildred who's the line worker becomes Mildred the robot trainer. She lifts her game. . . . We're not giving them tools that they have to go and study for years and years in order to use. They're tools that they can just learn how to operate in a few minutes.[61]

According to Brooks and several of the clients whose interviews are posted on the Rethink Robotics website (such as General Electric and Steelcase), Baxter does away with the prior association of industrial robots as "scary" entities taking away human jobs. Conceived of as the opposite of outsourcing logics of the past, this new iteration of human workers training robots to do the tasks that they had done before is disassociated from the numerous instances where laid-off (outsourced) workers trained their underpaid replacements and counterparts in the Global South.

Given the existence of software and hardware enabling workers to train a robot with no prior knowledge, the cobot raises the question of whether human factory and service workers have been further reduced to the pure body, while consciousness, intuition, and learning are now the purview of the robot inheritors of ostensibly dull, dirty, and dangerous jobs. This shift demands close attention to the parameters of the human–machine collaboration envisioned in the concept of "collaborative robots." For instance, even as the collaborative robot is positioned as the opposite of a robot that can take a worker's job as the robotic arms of the 1980s did, Brooks's perspective nonetheless predicts the extinction of factory workers, and thus falls in line with imaginaries of technology that project the end of labor- or class-based social movements. Mildred and other workers are not taught indispensable skills. In the above quote, Brooks emphasizes that no technical or engineering knowledge and expertise are required for an industrial worker to become a robot trainer in the twenty-first century. This means that "robot trainers" are themselves infinitely and quickly replicable. Additionally, the reader can infer that once the current (aging) generation of workers to which Mildred belongs dies out, they will already have trained robots to do tasks done by factory workers of the past. New (younger) workers will go on to do more exciting (skilled, creative) work, not the devalued work that their parents (like Mildred) want them to stay away from. Thus, in this new economy of people and things, it is robots that have vitality, whereas manual labor is situated in a necro-temporality that need only be managed for a short period of time. In this sense, Baxter reiterates the technoliberal futurity of a postlabor world.

Reimagining Sharing and the Commons

Though Marx's later writings consider technology as a tool for liberating humans from toil, for the most part, and against technoliberal arguments that technology is inherently revolutionary, technology increases toil and exploitation for workers. At the same time, technology is in fact responsible for the Marxian imaginary of labor as the material basis for asserting the uniform and revolutionary subject that will reclaim the commons appropriated by capitalists. For Marx and Engels, it is industrial machinery that creates the universalized quality of the labor-based proletarian subject in the first place by also increasing the unbearable quality of work, as

> owing to the extensive use of machinery, and to the division of labour, the work of the proletarians has lost all individual character, and, consequently, all charm for the workman. He becomes an appendage of the machine, and it is only the most simple, most monotonous, and most easily acquired knack, that is required of him. Hence, the cost of production of a workman is restricted, almost entirely, to the means of subsistence that he requires for maintenance, and for the propagation of his race. But the price of a commodity, and therefore also of labour, is equal to its cost of production. In proportion, therefore, as the repulsiveness of the work increases, the wage decreases. Nay more, in proportion as the use of machinery and division of labour increases, in the same proportion the burden of toil also increases, whether by prolongation of the working hours, by the increase of the work exacted in a given time or by increased speed of machinery, etc.[62]

Yet, as Haraway points out in the 1991 *Cyborg Manifesto*, universalizing categories intended to promote revolutionary consciousness negate histories of radical difference that must be acknowledged as they continue, both in human social life and in technoliberal design imaginaries, like those of the IoT and collaborative robotics that enact the surrogate human effect.[63] The *Cyborg Manifesto* offers a constructive critique of the limits of equating gender with gendered reproductive activity in materialist feminist arguments like Federici's, though it predates *Caliban and the Witch*. She notes that "one important route for reconstructing socialist-feminist politics is through theory and practice addressed to the social relations of science and technology, including crucially the systems of myth and meanings structuring our imagination."[64] This argument, made at a time when feminists were focused on discourses of the natural, and thereby not attending to technology, connects materialist feminism to technologies that are restructuring the social politics of difference.

Part of the call of the *Cyborg Manifesto* is for feminists to assert their influence and power by becoming the makers and designers of technology, and built into this call is an attempt to accommodate the inextricability of a multiplicity of situated feminisms with conversations in biology and biotechnology.[65] The manifesto does not assert a new technologically enabled commons, but rather experiments with imagining a shared feminist politics that embraces its entanglement with new technologies without asserting a unity among feminist politics nor a shared "common" of gendered activity, or what Haraway describes as "women's activity" of "reproduction in the socialist-feminist sense."[66] Here we find grounds to undermine the claims made by Italian post-autonomous thinkers, like Lazzarato, and Hardt and Negri, who imagine immaterial labor (knowledge and care work) as a commons under threat by capitalist appropriation, but also a commons that is available for all in service of anticapitalist revolution.[67] Haraway critiques the "unity of women [that] rests on an epistemology based on the ontological structure of 'labour.'"[68] Whereas Haraway is concerned with how socialist-feminists essentialize women's activity (reproduction) through analogy to labor, her critique also applies to an uncritical notion of the unity of the commons: "my complaint about socialist/Marxian standpoints is their unintended erasure of polyvocal, unassimilable, radical difference made visible in anti-colonial discourse and practice."[69] Haraway argues in favor of partial, rather that totalizing, explanations. She argues that new technologies, especially in biotech, reorganize production and reproduction, requiring a "feminist science."[70] She says, "What kind of constitutive role in the production of knowledge, imagination and practice can new groups doing science have? How can these groups be allied with progressive social and political movements? Might there be ways of developing feminist science/technology politics in alliance with anti-military science facility conversion action groups?"[71]

What can feminist makers of technologies that avoid unifying categories materialized through the surrogate human effect look like? Luis Martin-Cabrera brings together two transnational examples of anticolonial, perhaps feminist, engagements with technologies and the problem of universalizing the commons. He reads Alex Rivera's film *Sleep Dealer* as a futuristic musing that "establishes a dialogue with post-autonomous thinkers while exposing the limits of their assumptions by showing how technology and cognitive labor may actually reproduce forms of colonial exploitation and oppression rather than leading to automatic liberation from the shackles of physical labor. The film shows how technology is perfectly compatible with poverty and exploitation."[72] Specifically, he points out that the internal contradictions of capitalist forces of production will not lead to its demise, as per post-autonomous thinking, and for our purposes technoliberal thinking; rather, this can only result from "a political decision to struggle from within the system."[73] He then connects this to the *Ley Sinde*, a 2011 Spanish law that was defended as protecting the work of artists and their copyright by criminalizing web pages that, like *Public Library*, provide access to "illegal" archives.[74] Rather than embracing the preservation of nondigital property law being translated to the realm of the digital, or the freedom of information movement, both of which Martin-Cabrera rightly understands as reifying the digital, he offers a third position represented by the authors of a "Manifiesto en defense de una red libre." The authors insist on the enduring connection between the analogue and the digital, and therefore the immaterial and the material by linking this all to the struggle at the level of the Pueblo.[75] In other words, they refuse the erasure of the lives and labor of the people who continue to make the world, whether or not they are allowed visibility. As we have continued to argue in this and other chapters, invisibilizing labor has always been part of the surrogate effect that allows for the existence and seeming autonomy of the liberal subject. Universalizing the commons is therefore another project that serves the technoliberal instantiation of that project by asserting the digital as somehow immaterial, and the commons as once again only for that universalized liberal subject.

An overtly feminist imaginary of the use of new technologies and of generating feminist maker spaces is GynePunk, a Catalan-based feminist biohacking initiative. GynePunk is self-described as a "postcapitalist ecoindustrial colony" in Calafou, where they live and work on collectively owned property.[76] Located in the mountains outside Barcelona, this group lives and works on communally owned property, creating women's health technologies and preventive gynecological care kits for women who don't have access to primary care, and for themselves, toward their overall goal of "decoloniz[ing] the female body."

Their collective space includes a biohacking lab. Pechblenda, one part of an international open-source biology network called Hackteria, currently occupies the space, and conceived of GynePunk as an approach to "decolonize the female body" by way of developing accessible and mobile female gynecological and health tools.[77] Klau Kinky, one of the founders of GynePunk and designer of an emergency gynecologist kit for use by "immigrants without health coverage, for refugee camps . . . sex workers," and also members of the collective themselves, dedicated the project to Anarcha, Betsy, and Lucy, three women who were operated upon for fistula repair, without anesthesia, while enslaved by the renowned gynecologist J. Marion Sims in Alabama between 1945 and 1949.

Democratizing and liberating the instruments and protocols of obstetrics and gynecology is also part of the GynePunk project. The 3D-printable speculum is one of these tools whose design is open access with a Creative Commons license through websites like Thingiverse (figure 2.5).[78] It is circulated with web-based instructions for use, as well as directions to find further diagnostic and analytical tools on GynePunk's web archives. Three-dimensional printing, along with internet-based distribution of guides for use and complimentary practices of diagnosis and treatment, becomes a technology of both decolonizing bodies and health care, as well as communalizing knowledge. In this way, they engage not the political project of socialist feminists Haraway criticizes for universalizing "woman" through universalizing reproductive activity, but rather the political project Michelle Murphy has called "seizing the means of reproduction." By this, Murphy means "technically manipulating [the] very embodied relationship to sexed, lived being itself."[79] Both GynePunk and Murphy offer a direct intervention into Karl Marx's historical materialist theory of world

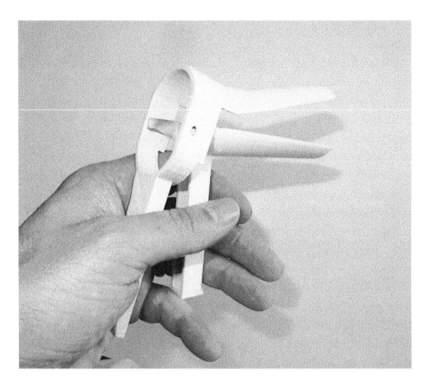

Figure 2.5. GynePunk speculum by gaudi, published June 5, 2015.

revolution, in which workers refuse to be instruments of industrial production, and instead socialize ownership of the infrastructure of industrial production, and therefore the economy. Murphy's concept of "sexed, lived being" offers a more expansive and multiple alternative to the revolutionary subject of Marx's labor power, disassembling the "sex/gender division by using technoscience to redistribute reproduction."[80] Unlike technoliberal imaginaries, where 3D printing is potentially revolutionary (bringing about the end of capitalism) because it might democratize the means of production, GynePunk's speculum, together with its distribution, manuals, support kits, and outreach and education, allows makers and users to "redistribute reproduction" and "manipulate their relation to sexed/lived being" as political projects that do not make claims to being revolutionary (which is revealed not to be just teleological but always threatening to flatten difference).

Both GynePunk's model of decolonizing the body through open-sourced blueprints for gynecological health technologies, like the 3D-printed speculum, and *Public Library*, with which we began, understand knowledge to be an essential part of providing common resources that do not assert a universal subject, body, or need, despite their inhabiting the same digital realm as that translated into private property by laws like Spain's *Ley Sinde*. What should be apparent after reviewing both the technoliberal imaginary of the new commons, exemplified in Rifkin's celebration of the IoT, as well as emergent formulations of the collaborative commons found as early as Haraway's *Cyborg Manifesto* and the work of GynePunk, are the structural inequalities that result from histories of colonial, racial, and gender dispossession that map directly onto new technological platforms. This also marks a continuing need to think about different kinds of collectives, and what they mean for the ways that the space and time of the political have shifted, even as the struggle to bring the knowledge and bodies that ethnic studies represents into the academy continues.

The impasse in linking technological futures to political collectivity, represented in this chapter by the conflation of revolution with techno-objects and the end of human labor (as the advent of full humanity for only the privileged few), calls for a dialogue between the multiple inheritances of socialism and alternate postcapitalist imaginaries in the present. J. K. Gibson-Graham points out that capitalism itself, imagined as a systemic concentration of power that transcended the revolutionary potential of socialism, is in fact "a set of economic practices scattered over a landscape."[81] How, then, do we understand the relationship between the afterlives of distinct, if interlinked, socialisms and the politics of revolution as a mode of social transformation?

Foregrounding the limitations of techno-revolutionary imaginaries as in fact colonizing the notions of sharing, the commons, and collaboration, as we have done in this chapter, we wish to excavate the past and present politics invested in collectives, in what is common and shared. Rather than assuming a Marxist, Leninist, or even immaterialist "commons" of affect and intellect, it is crucial to make legible specifically nonuniversal collectivist endeavors as connected to pasts that in some instances were overwritten by Cold War politics, and in other instances are self-generating communal investments in a common good. For instance, Gibson-Graham's theory of postcapitalism refuses a model of millennial future revolution and instead identifies models of local transformation, like that of the Zapatistas and the World Social Forum, which bring together movements of many political orientations as part of a global "movement of movements" that does not require transformation at larger scales.[82] They connect these "noncapitalist" local projects through a language of economic difference from capitalism rather than through a shared model of global transformation, and they argue that the subject of postcapitalist politics is marked by new ethical practices of becoming different kinds of economic subjects and beings.[83] *The Ends of Capitalism* "queer[ed] and deconstructed" capitalism, highlighting an affect within left revolutionary politics that focused "somewhat blankly" on a "millennial future revolution."[84] *Postcapitalist Politics*, the title and topic of their second book, refuses this blank focus of the left, and instead focuses on an emerging political imaginary that connects already existing localized economic alternatives to capitalism.

The ultimate project, then, is to continue taking into account the use of specific technologies in the service of racializing and devaluing particular populations and various modes of imperial domination, but at the same time, to be able to consider how those same sorts of technologies can be used to form or imagine different types of collectives beyond the nation-state, institutionalized religion, or class-based agitation. These collectives may emphasize both their continuity with historical forms, and also the importance of ongoing social justice imaginaries as they have been adapted to ever-emerging new social-technological platforms, political imaginaries that protest capitalist developmentalist and neoliberal cultural and economic projects.

3. Automation and the Invisible Service Function

Toward an "Artificial Artificial Intelligence"

The promotional video for the first social robot designed for home use, "Jibo," opens with a shot of a typical suburban home. The voiceover sets the scene: "This is your house, this is your car [the video cuts to a shot of a Mini Cooper parked outside the attached garage], this is your toothbrush [the video cuts to a shot of an electric toothbrush on a granite countertop], these are your things. But these [the camera zooms in on a family photo of what appears to be a white, heteronormative, nuclear family of adults and children of various ages] are the things that matter. And somewhere in between is this guy [the camera pans to an eighteen-inch-tall, white, two-piece object that rotates a spherical display with a flat, black, round screen displaying an illuminated greenish orb]. Introducing Jibo, the world's first family robot. Say 'Hi,' Jibo."[1] It responds, "Hi Jibo," in a surprisingly nonrobotic cartoonish voice, then giggles and squints the green orb on its display into a shape reminiscent of a human eye on a smiling face (figure 3.1). It is immediately clear that it is through altering its physical "expression" to reflect an affective state that Jibo interacts with humans. It performs attention by using a three-axis motorized system to point its screen in the direction of a speaker and to perform small dips and bobs that indicate a deferential focus on the speaker. Its speech engagement performs an obeyant interest in the commands, needs, and affective states of that speaker.

Released by roboticist Cynthia Breazeal to promote her new domestic robotics company in July 2014, this video was intended to encourage preorders of Jibo to raise capital to fund its further development. Following this initial scene, the video continues by moving through a set of familiar heterosexual middle-class nuclear-family scenes. The narrator describes Jibo as the scenes unfold. First is a children's birthday party, where we hear that "he's the world's best cameraman. By tracking the action around him, he can independently take video and photos, so that you can put down your camera and be a part of the scene." We then move to the kitchen, where an older family member up to her elbows in flour while kneading dough is told that her daughter has sent her a text message. We hear Jibo addressing her by first name after first politely saying, "Excuse me," and the narrator tells us that Jibo is also a hands-free communication device. Finally, as we see a scene of children getting ready to sleep, we hear: "He's an entertainer and educator. Through interactive applications, Jibo can teach." We view a scene of Jibo playing a recording of a children's story at bedtime, showing simple corresponding images on its screen. Jibo's body moves continuously throughout its scenes, denoting its attention and interactivity. When the screen gives us an occasional perspectival shot from Jibo's vantage point, we see the same family scenes with people's faces framed and named by facial recognition software. We are then told that "his skills keep expanding. He'll be able to connect to your home." After reviewing Jibo's genealogy through a montage of popular cultural images of friendly interactive robots, including "Rosie" the robot maid from the animated *Jetsons* TV program, and the escapee military research robot "Johnny 5" from the film *Short Circuit*, we are told that Jibo is "not just a metal shell, nor is he just a three-axis motor system. He's not even just a connected device. He's one of the family."[2]

Jibo is a smart device designed for home use and exists on a continuum with Amazon's popular Alexa smart entertainment hub, but it is claimed as a "robot" by Brazeal, ostensibly heralding a new consumer frontier. This commercialization of social robots, the first of their kind, inserts them into the home in a way that normalizes the white US middle-class nuclear family form. More specifically, it advances the norm of the household as an autonomous and self-sufficient economic unit, even as it proves that unit to

Figure 3.1. Jibo interacting with the idealized image of the white heteronormative family. Promotional still from the Jibo press kit.

require substantial and invisible support labor. How is Jibo's embodiment
enacting a sociality more complex than would, say, a smartphone with a
rotating head? What import does the affect and embodiment of an obeyant physicality and machine subjectivity have, given the goal of making a
domestic social robot as commonplace as a car or a toothbrush? The engineering imaginary behind Jibo's emotional interactivity and obeyant attention brings the history of robotics into the immediate material context of
the gendered and raced history of domesticity and of the normative family
form in the post–World War II United States.

The "robot" Jibo, also described as a platform, represents a technological
update to the domestic architecture that supports the seeming autonomy
of the liberal subject. Here, technoliberalism reinscribes a racialized and
gendered service labor infrastructure that derives from the domestic division of labor under colonialism. It functions as an architectural update
to the hidden passageways and serving spaces in US antebellum residen-
tial architecture that were meant to hide the enslaved "hands that served"
and yield the service without any sociality between those served and the
servants.[3] As with the development of technologies promoting outsourced
service industries in India built on a design imperative of importing the
labor but not the lives of workers, as described in Vora's book *Life Support*,

Jibo is designed to preserve the architecture of erased service work that allows the nuclear hetero-patriarchal family economic unit to continue.

In this chapter we address how present-day disappearances of human bodies take place in the information milieu, specifically through platforms intended to disguise human labor as machine labor. Whereas Jibo, as a physical robot, can be brought into the center of the hetero-nuclear household without disrupting the erasure of human domestic and service work that supports the reproduction of the whiteness and patriarchy of this family form, the information milieu must rely on a rearrangement of space to uphold the myth of human-free labor under technoliberalism. The racial grammar of the surrogate effect is here tethered to the programming architectures of virtual spaces that erase the sociality of laboring bodies. These disappearances are a part of technoliberal innovation related to the racial dynamics of collaborative robotics and the sharing economy that we addressed in chapter 2, even as they introduce a different aspect of the fantasy of human-free labor. Thus, we might ask, what is the relationship between a techno-futurity in which the human is engulfed into an Internet of Things, and a present in which human labor continues to be irreplaceable even as it is hidden beneath the fantasy of what AI can accomplish? Concealing the human worker as the source of information processing, data collection, and service work has become a central feature that enables the conception of the fourth industrial revolution and the second machine age as the socioeconomic paradigm shifts. Why must the worker be concealed to enable the growth of the digital economy? Design projects that hide service labor advance the project of technoliberalism by contributing to the seeming inevitability of the domestic realm as an atomized and apparently autonomous economy where the support of life is an individual, rather than a social, concern.

Analyzing platforms including the Alfred Club service website and Amazon Mechanical Turk (AMT), a data management service, we turn our attention to how humans themselves are performing the work of technologies that are claimed to replace the need for human workers. This is what AMT has framed, tongue in cheek, as *artificial* artificial intelligence. Our emphasis on the disappearance of human labor allows us to theorize surrogate humanity as not just about a set of new technologies, but more importantly, about how the fantasy of human-free social environments, including everything from cyberspace to the domestic sphere, is concerned with replacing the racialized and gendered surrogates enabling freedom for the universalized liberal figure of the human with technological surrogates. In other words, we argue that technologies that erase human workers are designed to perform the surrogate effect for consumers, who consume the reassurance of their own humanity along with the service offered.[4] The surrogate effect can command capital, as well, when venture capitalists who fetishize automation and digitization see apps that offer a technological veneer to what are in fact long-standing human services.[5]

Platforms like AMT and Alfred Club perform the surrogate effect by affirming the humanity and subjecthood of their users in ways that both rehearse and innovate upon the prior racial and gendered politics of labor. Put otherwise, emerging technologies like AMT and Alfred Club simultaneously exploit gendered and effaced service work and demand that the worker participate in effacing herself as a subject.[6] Drawing attention to the surrogate effect produced by the socio-spatial dynamics of distancing and erasure of service workers within service platforms enables us to center questions of racialized and gendered difference where they otherwise may be displaced or obscured in the postracial fictions of technoliberalism. Analysis of the surrogate effect thus requires a feminist, critical race and postcolonial science studies approach to the field of labor politics. The examples in this chapter show how the surrogate effect delimits what counts as work, and what counts as a valued social relation, because it defines those who count as recognizable subjects in those areas.

The Butler as Transparent Commodity

As a hyperactive medium of free labor generating and circulating social and economic value within late capitalism, the internet, which has given rise to the sharing economy and imaginaries of servitude that are quite different from that of humanoid robots, necessitates an investigation of how racialized and gendered imaginaries of freedom and labor have shifted, as well as what the implications of such shifts might be.[7] For instance, late capitalism does not just appropriate creative/unalienated labor, but rather nurtures, sustains, and exhausts it.[8] In fact, the "free" quality of those excess productive activities that internet users undertake with pleasure even as they are mined for value, such as clicking "like" on a Facebook page, or clicking on advertisements linked to browsing activities, "puts the desire for creative work into tension with the information economy's demand for knowledge as added value."[9] While shifting the relationship between producer and consumer, according to Tiziana Terranova, the commodity nevertheless does not fully disappear in this arrangement. Instead, it becomes "more transparent" as it showcases "the labour of the designers and programmers," with continually changing content that draws users back.[10]

This conception of transparency, however, doesn't allow for an analysis of "noninnovative" human labor, or service labor including that organized through various apps such as Uber and TaskRabbit, as a commodity within the digital economy. Instead, the transparency of service workers as commodities managed through such platforms resides in their organization as those who quite literally are meant to go unseen by consumers. Racialized and gendered labor has long been subsumed within and then disappeared from technological infrastructures. The contemporary conditions of labor made visible and invisible by so-called smart infrastructures are the inheritors of the gendered and racialized invisibility of those we might consider early tech workers. Jennifer Light and later Wendy Chun have discussed the early history of computing when "computer" was a clerical job title for positions held by women who computed calculations as a devalued and gendered form of labor.[11] Chun reflects on the later history of software development out of this history, noting that "One could say that programming became programming and software became software when commands shifted from commanding a 'girl' to commanding a machine."[12] In a different geopolitical and historical context, Lisa Nakamura has addressed how from 1965 to 1975 the Fairchild Corporation's semiconductor division operated a large integrated circuit manufacturing plant in Shiprock, New Mexico, on a Navajo reservation.[13] Nakamura considers the racial and gendered architectures of electronics manufacturing, emphasizing the need to think about what she calls "insourcing" (as opposed to outsourcing, which is generally associated with tech work). Indeed, during the time period covered by her research, the Fairchild Corporation was the largest private employer of indigenous workers in the US. The circuits, produced almost entirely by female Navajo workers, were used in numerous technological devices such as calculators, missile guidance systems, and other early computers. Because of their weaving skill (part of the "traditional" Navajo arts), Navajo women were represented as "naturally" suited to the detailed labor involved in building the circuits. Yet, as Nakamura explains, Native American women have been all but erased from the official histories of the microchip and its exponentially growing capacities that enable the contemporary economic transformation.

In the present day, the reemergence of racialized and gendered workers in and through ostensibly transparent commodities (ones that would make the labor that produces them visible) can be understood as racialized, not just in the earlier ways in which women *were* computers or Navajo women created circuits; rather, the transparency of the commodity as dull or dirty labor is also racialized in the ways in which it affirms a particular notion of human freedom, leisure, and happiness. As an example of service workers who function as transparent commodities to enable the experience of full humanity by producing the surrogate effect for users, we consider Alfred Club, the 2014 winner of the blog TechCrunch's annual competition for tech start-ups, which was hyped as an innovator in the realm of the emergent tech-driven sharing economy.[14] TechCrunch touted Alfred as innovative because it is "the first service layer on the shared economy that manages your routine across multiple on-demand and local services."[15] "Alfreds" are individual workers made invisible and interchangeable through the Alfred Club's platform; however, every aspect of the Alfred service advertises the possibility of zero personal interaction between the subscriber, the person who is to act as the butler (or "Alfred"), and the additional service providers whom "Alfred" manages. As the company boasts, "Alfred is an automatic, hands-off service that hums along quietly in the background of your life—so you can be free to live yours."[16]

Unlike robots such as Jibo, whose material presence and visibility within the home space is the point of its technological innovation (even if Jibo is no more innovative in its function than a smartphone, the idea of a robot in the home creates the fantasy of a future where robots do work for us), the point of Alfreds who work within the digital economy is to be invisible. The innovation is the interface (the user interacts with a platform rather than a person), and thus the platform enables the fantasy that technology is performing the labor, though in fact it is being done by human beings. The interaction on the Alfred platform with the most potential for personal contact occurs at the very start of service. When a person signs up for service through the company's website, she is assigned an Alfred. The subscriber receives access to a picture and background check of her designated Alfred as the primary individual to whom she can then begin to delegate tasks, such as buying groceries, picking up the dry cleaning, and cleaning the house. This is also the person who initially picks up the keys to the subscriber's home, though even this level of interaction can be bypassed if the subscriber chooses to send a picture of her keys so that the company can make a copy of them. Indeed, this service platform's primary innovation and product is the erasure of contact between service workers and subscribers. For instance, the Alfred service at times uses some existing on-demand services, such as Handy for cleaning and TaskRabbit for errands, but with Alfred, the subscriber need not be home when the task happens.[17] As critics of the web service point out, what distinguishes Alfred from a non-high-tech personal servant is the depersonalizing aspect (all Alfreds even share the same name, as the subscriber need not remember their assigned Alfred's actual name), and the cost: just $99 per month.

With this emphasis on disappearing the service worker, it is possible to read Alfred as targeting the customer who would be uncomfortable having a regular personal assistant or servant, and who, perhaps espousing a (techno)liberal ethos, wants to make sure that the Alfreds are not underpaid or abused. For instance, tech writer Sarah Kesser describes her concerns as she decided to give the service a try:

> Am I really so lazy that I can't even lift my thumbs to my own iPhone to ask SOMEONE ELSE to clean my home or do my laundry? Am I contributing to unfair labor practices—like those for which workers recently sued Handy—or at the least, to the next startup bubble, by encouraging this startup nesting doll of a service? But Alfred has one argument that is hard to deflect: Returning to my home after work to find all of my errands completed, without any effort on my behalf, sounds amazing. Despite my qualms, when they offer me a pre-launch trial, I'm in.[18]

Liberal concerns surrounding labor practices and the exploitation of labor can thus quickly give way to the pleasures and enjoyment of services precisely because Alfreds are successful if they completely erase the signs of their presence (one magically finds one's errands are complete upon returning home). Invisibility makes it possible to remove (and move beyond) initial concerns about the physical bodies tasked to pick up subscribers' dirty laundry. Put otherwise, though Alfreds are actual workers, their function within the Alfred platform is to act on behalf of technology that will ostensibly one day be doing this work. Their labor, which is invisibilized and invisibilizing of other service providers as they complete the busy professionals' onerous chores, is necessary to the functioning of adult life.

Alfred founders Marcella Sapone and Jess Beck build a technoliberal ethos of unseeing dull and dirty work as a precondition for happiness into the very origin story of their company. According to the Alfred website: "One fateful night after a long, demanding day at work, [Marcella and Jess] ran into each other in the laundry room of their apartment building. The overwhelming sentiment? Frustration. Leisure time shouldn't be a luxury; it should be a right and a reward for working as hard as both of these women did (and still do!). A pact was made: No longer would they let mundane chores control their lives."[19] This is framed in a liberal feminist rhetoric, and we find both the acknowledgment and the refusal to recognize that the work that Alfred makes invisible to the subscriber is work that has always been invisible: women's work. The right to leisure time for "these women" because of their hard work inspires a pseudo-feminist pact: mundane work is not the domain of such women as themselves. The narrative simultaneously refuses the long history of black, Asian American, and Latina women who have historically performed domestic work in white homes.[20] This liberal feminist narrative does not include the rights and rewards for people finding employment through their service platform. This brand of feminist liberation marks an essential quality of the liberal human: like the founders of Alfred, who are white-collar professionals, "women" and racial others signifying liberal progress within technoliberalism are now elevated into a category of people who deserve happiness that relies on someone else's invisible work.[21] In other words, in this narrative we find one instantiation of the surrogate effect, in which the experience of full humanity as a subjective, but also affective, state is possible only through the relation to a diminished subject. The Alfred service platform thus both inherits the prior forms of racialized and feminized, intimate labors supporting the nuclear, heteronormative, white family form and disappears such intimate service obligations (and possible annoyances) from the modern home in a mode consistent with notions of racial and sexual progress. Those who enter your home, touch your dirty clothes, and clean up after your "epic parties" know your habits and preferences for baked goods, but you need not see or know them.[22]

As the origin story of the service elaborates, "Since [their initial meeting], Marcela and Jess have worked tirelessly to create the first ever non-intrusive, recurring, in-home service that virtually everyone can use. With Alfred, these innovative women have ensured they personally will always come home happy—and they want the same for you."[23] This reconfiguration of human intimacy and intimate spaces through the human–machine interface that invisibilizes service work is not only haunted by the heteronormative nuclear family and the asymmetries in how the work that has kept this formation functional is differentially valued; it is also haunted by the imperial-racial production of intimacy as often violent.[24] Thus, while the race, gender, sexual orientation, or domestic arrangements of the Alfred subscribers (who are implicitly diverse) necessitate the reimagining of intimate laborers as anonymous workers, no longer bound by their racialized or gendered bodies, such service platforms also inherit prior (imperial) imaginaries of happiness and freedom that were relationally consolidated and unevenly distributed.

The Alfred service's emphasis on worker invisibility in completing the chores of busy professionals and enabling their happiness, as well as Alfred's "innovation" of coordinating among other services in the sharing economy, align Alfred with the ethos of disappearing degraded work in the object-centered universe of the Internet of Things. People (Alfreds) and artifacts in the Internet of Things could be said to perform a similar function of enchantment in techno-imaginaries of the sorts of innovations that will maximize human happiness and freedom. According to *Wired* magazine, in the new world of programmable objects that form the Internet of Things, "we are seeing the dawn of an era when the most mundane items in our lives can talk wirelessly among themselves, performing tasks on command. . . . Imagine a factory where every machine, every room, feeds back information to solve problems on the production line."[25] Positioning its platform as part of this enchanted techno-future, a cofounder of Alfred noted that with her app, "the services in your life know you and they are automatic."[26]

The vision of a future in which the intimacy and relationality of consuming human service is erased through automation depends profoundly on a present in which devalued subjects performing miserable (and historically classed, racialized, and gendered) forms of labor are considered replaceable by technology. Marx's original critique of industrial capitalism's dehumanizing effect on workers can still be applied to techno-utopic fantasies of a postcapitalist, posthuman world: "The bourgeoisie cannot exist without constantly revolutionising the instruments of production, and thereby the relations of production, and with them the whole relations of society. . . . Owing to the extensive use of machinery, and to the division of labour, the work of the proletarians has lost all individual character, and, consequently, all charm for the workman. He becomes an appendage of the machine, and it is only the most simple, most monotonous, and most easily acquired knack, that is required of him."[27] Fantasies of a networked world in which automation, programmability, and data engender unprecedented freedom in fact reproduce, and perhaps even expand, the sphere of devalued labor, transposing categories of abjection from the human into the very infrastructure upon which the techno-utopic future depends.

That multiple Alfreds, who are actual workers rendered as "automatic" services, facilitate daily life for privileged subscribers reveals how the disappearance of bodies historically marked as "things" enables the fantasy of enchanted techno-objects that populate the sharing economy and IoT, as well as the hope that these objects will free those fully human from the efforts of mundane labor. The sharing economy of programmable "objects" and service platforms that utilize people only to make them invisible (reducing them to their functionality) enacts the surrogate effect, reinscribing differential relationships to the category of the human, now mediated through digital networks and informatics.

The Artificial Artificial Intelligence
of Amazon Mechanical Turk

Digital service-based companies like Alfred, or the goliath Amazon, show how digital platforms engender new relations between human labor, space (where people who perform needed tasks are located and how they can be located by those who need their services), and finally the product itself (that is, the service, and what Amazon CEO Jeff Bezos calls "humans-as-a-service").[28] For instance, the experience of digitized, near-instant delivery that is part of Amazon's basic warehousing strategy, essential to its commercial success and as innovative as Sam Walton's Walmart infrastructure was for brick-and-mortar commodity sales, relies on subjecting warehouse stock workers to dehumanizing work shifts and physical demands for speed and endurance required to move products rapidly enough to achieve same-day and next-day delivery. The end product of Walmart's warehousing infrastructure was faster and cheaper delivery of goods, which required more and more human work hours to achieve. Walmart's warehousing infrastructure, or even the updated model used by Amazon, which speeds up the labor of both sorting and delivery, as well as adding personal delivery in a shorter time frame, requires many anonymized hands to contribute to the end product. The increased speed of delivery and decreased prices are subsidized by a greater and greater number of workers who put in long hours beyond the forty-hour workweek.

Crowdsourcing platforms operate in a similar way, but because the platform creates a veneer of machine intelligence, the increased labor exploitation involved is partially obscured. Together, the similarities between warehousing and delivery infrastructures and crowdsourced labor infrastructures undermine technoliberal aspirations that project the obsolescence of human labor in the postindustrial economies of the post–Cold War world. Of course, the necessary labor has not only failed to decrease— it has increased even as lower compensation is justified through a rhetoric that crowdsourced microtask work is flexible, voluntary, and provides "extra" income that supplements workers' primary wages.

The Amazon Mechanical Turk Internet Marketplace (AMT) is just such a large-scale crowdsourcing software platform where employers can post "human intelligence tasks" with a price for each task.[29] In the world of coding and online retail, programmers continue to struggle to come up with the holy grail of artificial intelligence—software that can use algorithms to categorize the seemingly infinite amounts of data, including user-generated data (personal photos, rants in the blogosphere, kitten humor), into value-producing data sets for e-commerce. AMT was originally designed in 2005 to solve a problem very difficult for artificial intelligence, namely identifying duplicate product pages among the millions Amazon has created for its products.[30] This task is very simple for most human workers. It was designed to provide intellectual piecework to employers, called "requesters," on a per-task fee system; individual workers, who often self-refer as "turkers," can pick up a human intelligence task, or HIT, through the platform.

After creating a profile and registering with AMT, a "turker" can choose among HITS categorized by the nature of the task, the difficulty or skill level required, and the amount of compensation. Beneath a simple flow chart that describes the process as: "Find an interesting task, work, earn money," AMT's splash page offers a familiar Amazon yellow shopping button to click and "Find HITS now." What follows is a list of tasks with four- to ten-word descriptions such as "decide the general category of a video," "describe the setting of a short animation clip," or "transcribe up to 35 seconds of media to text." These tasks have an amount of time allotted (five minutes, fifteen minutes, etc.) as well as a "reward" paid per task (most often two to ten cents). People who want to hire turkers will find a similarly oversimplified flow chart: "fund your account," "load your tasks," and finally, "get results." The same yellow shopping button on the requester side of the splash page offers "get started." This button leads to a more complicated page than that for turkers, as it offers tabs for creating projects, managing existing projects, and developer options, as well as embedded tabs for pricing, partnering, and case studies of projects that have used the platform's humans-as-a-service. For their third-party service, Amazon takes a percentage from the employer.

When a huge amount of data needs to be organized into culturally legible categories, like classifying pornographic images or linking items for the "similar products" function on retail sites like Amazon, this work is sourced task by task to a global pool of workers and performed as microtasks. The internet reaches all the world's time zones, and so, as Lilly Irani has provocatively proposed, the sun never sets on Amazon's technology platform.[31] At the same time, AMT crowdsourcing utilizes mainly workers from the US (80 percent), with 20 percent of workers coming from India.[32] There is considerable variation in the distribution of race and gender between different tasks and times of day that tasks are completed, and, whereas before 2013 a majority of turkers were female, since 2013 there has been a rough balance between male and female turkers.[33] Tasks are similarly varied. An estimated 40 percent of AMT HITs involve creating and disseminating spam through social media and other outlets.[34] Companies use turkers to look up foreign postal codes, transcribe podcasts, match websites to relevant search terms, categorize images based on their subject, and flag objectionable content on websites.[35] Most recently, as "computers become smarter," the so-called human cloud's tasks have shifted, and now include training computers to be more "humanlike." Examples include human turkers selecting similar photos to enable the artificial intelligence of sites like Pinterest "to get better at predicting the pins a user will like."[36]

The surrogate effect of AMT masks the fact that artificial intelligence does not yet exist without the human. Tellingly, AMT is named for the Mechanical Turk, an eighteenth-century chess-playing automaton that toured around Europe and defeated prominent members of society at chess (figure 3.2). In reality, a chess master was hidden inside. Both the historical Mechanical Turk and the platform are *artificial* artificial intelligence, replicating the imagined speed of digital processors by giving employers access to an

Figure 3.2. The mechanical chess-playing Turk, a parlor trick of the
Austrian imperial court in the eighteenth century, from which Amazon
Mechanical Turk takes its name. © Interfoto/Alamy.

enormous pool of temporary workers for high-data tasks when there isn't
time to develop appropriate algorithms to do the job. As the person inside
the original eighteenth-century Mechanical Turk, the human is subsumed
so that the fantasy of technological enchantment may continue.

In this connection, AMT functions like a dress rehearsal for an artificial
intelligence to come. The work that artificial artificial intelligence does
to propagate technoliberalism is to perform a future predetermined by
contemporary imaginaries of the kind of work that can most usefully and
productively be automated for the purposes of capitalist acceleration and
accumulation. Susan Buck-Morss has written about this kind of dress rehearsal in a different context—that of the early years of the Soviet Union.
Buck-Morss contends that unlike in the capitalist countries where industrialization was already a catastrophic reality by the 1910s and 1920s, in the
Soviet Union this epoch was still a dreamworld.[37]

> The cult of the machine preceded the machines themselves. . . .
> Under the pretechnological conditions that existed in the early So-
> viet Union, in contrast, the cult of the human-as-machine sustained
> a utopian meaning. Its ecstatic intensity in the 1920s, at a time when

the factory workforce had disintegrated and the country was strug-gling simply to restore the pre–World War I industrial capacity, *an-ticipated mechanized processes* rather than being a defensive response to them. The Central Institute of Labor . . . founded in 1920 to im-plement Taylorist work methods imported from the United States, was run by a poet. *It was an experimental laboratory in the mechanized rhythmics of labor.*[38]

Following Buck-Morss, service platforms like AMT can be read as labora-tories for technoliberal futurity, faking the disappearance of human labor until this can become a reality. GrubHub, an online food-ordering platform and app through which users can get deliveries from restaurants that do not provide delivery service themselves, posted a billboard in the fall of 2017 by the Bay Bridge that connects San Francisco and Oakland, making visible this very process. With a drawing of a robot on the side, the text of the billboard reads: "Because self-delivering food is still in beta: Grub-Hub." Food delivery companies like GrubHub, which are periodically in the news as developing automated delivery through the use of drones and self-driving vehicles, are still relying on a pool of precarious and exploit-able laborers who can meet the needs of the on-demand present enabled by mobile devices.[39] The envisioned future must first further devalue work such as food delivery before it can render it obsolete.

With the example of AMT, the historical gendering of what is deemed the noncreative and noninnovative work required of the turkers carries forward the negative valuation of reproductive work from the hetero-patriarchal domestic sphere. "Wages on the platform, where workers are paid per task rather than per hour, are usually below the US federal mini-mum wage of $7.25 per hour. [A] Pew survey found about half of Turk-ers make less than $5 per hour. Nearly two-thirds of the tasks posted on the site pay 10 cents or less."[40] The justification for the low wages is that the tasks are framed as earning turkers pin money, or even as entertain-ment. An article in *Salon* magazine that discussed the work of AMT came out shortly after the platform was launched, with the headline, "I Make $1.45 a Week and I Love It!"[41] Discussing the popularity of HITs like trivia questions posted by sports bar patrons, where an argument about a sports history fact can be submitted to an online service that will give you the answer by posting an AMT HIT, the article explores the experience of a few turkers who work as an alternative to "just sitting there." Like the feminized subjects performing domestic work and childcare in the hetero-patriarchal private sphere, who were the result of a historical process that created a new subject, the housewife, the turkers take HITS as a form of productive work not recognized as labor in the private sphere. The housewife, like these turkers, also ostensibly did that work out of love. According to Maria Mies, the nature of that love was also a historical construct.[42] The affect of leisure and nonwork attached to casual AMT work is part of the system that disguises the digital economy's dependence on human labor.

The low wages for the surprisingly highly educated pool of casual laborers are exacerbated by the fact that Amazon acts as a third-party organizer for Mechanical Turk, who lets employers determine how and even whether to compensate workers, and hosts various forms of training and evaluative software processes that create workers appropriate for the task at hand. However, there is no screening or evaluation process for workers to choose or evaluate employers, and therefore they cannot know how reliable their employer is at actually paying the promised price, or whether the task matches its description. One news article describes how "Online forums are filled with workers complaining about how difficult it is to communicate with the people assigning the tasks. Referring to employers, one turker says, 'When you can't communicate . . . you lose empathy, and when you lose the empathy people are devalued.'"[43]

The algorithmic managing of labor that happens through crowdsourcing platforms like AMT, but also through applications like GrubHub, Uber and Lyft, or Handy, has merged the feeling of freedom and entrepreneurial resourcefulness with the tight logics of expedience, efficiency, and maximum profit. Alex Rosenblatt has shown, through a three-year-long ethnographic study of drivers working for Uber, that these workers actually face more limited choices about how to conduct their business than taxi drivers, while also being coerced to work longer hours through the interface of the Uber application.[44] Whereas in the world of programming the challenge is to create an intelligence that can pass the Turing test, or other tests of human likeness, algorithms eliminate communication, streamlining how skills are tested and how they indicate reliability. As Irani has argued with regard to AMT, this "preclude[s] individually accountable relations."[45]

Data-scrubbing labor, or "commercial content moderation," which allows for image filters in social media sites like Facebook and search engines like Google to maintain community standards of acceptable content, offers another example of a growing site of the low-paid and miserable work of producing the timely and sanitized digital experience internet users have been groomed to expect and demand. This work is sourced through intermediary firms like TaskUs to business process outsourcing units in the Philippines.[46] As with Mechanical Turk, employers not only have intellectual property rights to the products of microwork, but they can refuse to pay for work if they deem it of insufficient quality.

By hiding the labor and rendering it manageable through computing code, human computation platforms have generated an industry of start-ups claiming to be the future of data. Hiding the labor is key to how these start-ups are valued by investors, and thus key to the speculative but real winnings of entrepreneurs. AMT is thus exemplary of a new iteration of outsourcing—here the tasks that make the higher-value work of creative and innovative laborers possible—but it does so in a manner that does not unsettle the comfort of people accustomed to the pace of digitized feedback. AMT turkers get a task done as well as participating in the necessary occlusion of their subject position in order to provide the effect of being artificial labor and artificial intelligence.

Hacking Turk: Challenging the Invisibilizing Logics of AI

The ironies and assumptions tied to fantasies of artificial artificial intelligence, and the kinds of necessary disappearances that undergird technoliberal dress rehearsals for a technoliberal postlabor future, have been the subject of several critical and creative works. A US-based digital artist, Aaron Koblin, created a project called "sheep market" that highlights the ways that intellectual property law, and the racialized and gendered history of who counts as an innovator, a creator, or an author, is intimately connected to who remains the invisible laborer who produces the conditions of possibility for that work. Koblin posted a HIT on AMT that looked like the top of figure 3.3. He offered two cents per sheep drawing, which the turker would complete using the built-in drawing tool. Koblin then assembled the ten thousand resulting sheep drawings on his website (bottom of figure 3.3; detail in figure 3.4), where you can mouse over an individual sheep, and if you click on it, you can view the playback of the recorded process of drawing that resulted in the sheep, allowing the viewer to consume

THE SHEEP MARKET

10,000 sheep created
by online workers.
More...

Figure 3.3. "TheSheepMarket.com is a collection of 10,000 sheep made
by workers on Amazon's Mechanical Turk. Workers were paid 0.02 ($USD)
to 'draw a sheep facing to the left.' Animations of each sheep's creation
may be viewed at TheSheepMarket.com."

the labor process as well as the result (see figure 3.5). He has also set up a
gallery installation version of this website.

Koblin then offered blocks of twenty drawings from the complete set
for sale for twenty dollars. He notified the workers, who had a variety of
responses, and this created some dialogue about the ramifications of the
AMT system. The first turker discussion thread was entitled "They are sell-
ing our sheep!" asking, "Does anyone remember signing over the rights to

Figure 3.4. "'If you please—draw me a sheep. . . .' When a mystery is too overpowering, one dare not disobey. Absurd as it might seem to me, a thousand miles from any human habitation and in danger of death, I took out of my pocket a sheet of paper and my fountain-pen. But then I remembered how my studies had been concentrated on geography, history, arithmetic and grammar, and I told the little chap (a little crossly, too) that I did not know how to draw. He answered me: 'That doesn't matter. Draw me a sheep.'" *Le Petit Prince*. From sheepmarket.com.

What's going on here?

The sheep market is a collection of 10,000 sheep created by workers on Amazon's Mechanical Turk. Each worker was paid $.02 (US) to "draw a sheep facing left."

Average time spent drawing each sheep: 105 seconds

Average wage: $0.69 / hour

Rejected sheep: 662

Collection Period: 40 Days

Collection Rate: about 11 sheep/hour

Unique IP addresses: 7599

Figure 3.5. Statistics from sheepmarket.com.

our drawings? Someone should contact them and see how much they'd charge you to buy back the rights to one of your own sheep."[47] The truth was, the turkers had no legal rights to anything they produced through the HIT, and as of late 2016, no one has taken action to see how they might sue for rights.

As the Sheep Market project begins to illustrate, authorship, invention, and innovation are regulated as knowledge commodities in the corporation-driven market. This primary understanding of creative authorship is gendered, and relies upon a distinction between authorial, masculine discovery and feminine, reproductive, servile support labor. In biological and biotechnological research and development, and in genetic therapy research and development in particular, there are very specific notions of what counts as invention, and these are engaged with legal protections of intellectual patent and property. The history of scientific invention and experiment is characterized by a privatized system in which a privileged investigator, almost always an independently wealthy man, was legally and officially the author of the discoveries that came out of the work of hired technicians in the lab, "because of [his] resources and class standing."[48] Laws protecting intellectual property rely upon historically gendered notions of active versus passive creativity, where "support" labor, like that performed by nonauthorial lower-class hired workers or embodied or physical production, does not figure as producing property, and is therefore not recognized as an invention or the result of creative labor.[49]

Whereas the Sheep Market project troubles technoliberal distinctions between the creative labor that cannot be automated and noncreative devalued work destined for obsolescence with improvements in AI, another project disrupting the logics of invisibilized and devalued labor propagated by platforms like AMT is "Turkopticon," a project of Lilly Irani and Six Silberman. This is software that potentially gives turkers a way to create a database of feedback on employers that would help them determine the reliability of pay and the accuracy of their task descriptions.[50] One part of this intervention is a website where turkers can post and read reviews of requesters, who, as mentioned earlier, do not have evaluative statistics recorded by the AMT platform, as workers do. Another part of the project includes an app that a turker can add to the Chrome browser that will attach these reviews of requesters to their posted HITs once it is installed, if any reviews exist. While this isn't collective action in the tradition of "Workers of the world, unite!" through the work of the physically isolated single programmer (or, occasionally, small group hackathon), it engages the concern of organizing data as information to empower crowdsourced data piece-workers. This project is thus not only of interest because it turns novel, exploitative technological platforms against themselves, but because it is doing so as part of a socialist ethic of workers' rights. It thus represents an important new political terrain tying software-based social interfaces to political subjects (the online pieceworker, hacker, or dissident program-mer) and to new kinds of political consciousness (an individual creating a platform that may or may not be taken up by the constituents she perceives as being in need of it). Spontaneous social formations that arise through the technologies and conditions of production themselves can be read as temporary affiliations or investments between subjects who are accidentally connected through the irreducible heterogeneity of use-value carried in the commodity.[51] They can also be read as new political forms of social life, and as reminders of the political importance of the fact that "we are not the subjects of or the subject formations of the capitalist world-system. It is merely one condition of our being."[52]

The Sheep Market and the Turkopticon trouble the value placed on the surrogate effect through the proliferation of platforms like AMT and Alfred that disappear the actual human labor at work behind artificial intelligence. They demonstrate how technoliberal fantasies produce a product of seemingly human-free services that reassures the consumer of their own humanity, even as that product, at the same time, demands that workers use their labor to erase their presence. The labor of effacing one's presence, however, doesn't count as work on these platforms. Understanding technologies that demand that the worker erase their very existence as enacting the surrogate effect requires us to shift our definition of human–technology interaction. We argue, therefore, that analyzing technologies that exploit while erasing labor through the lens of the surrogate effect is a feminist, antiracist, and uncolonizing approach to the problem of the ongoing production of the post-Enlightenment human as a racial and gendered project in a technoliberal epoch that is touted as being postrace and postgender.

4. The Surrogate Human Affect

The Racial Programming of Robot Emotion

"The Robots Are Anxious about Taking Your Job."[1] So reads the 2016 headline of a news article explaining that as robots increasingly make their way into the everyday lives (and homes) of most Americans, humans will need to begin to cope with the unprecedented technological advances of the present day. Robotics expert Joanne Pransky, who calls herself the "first robotic psychiatrist," is one of the figures in robotics featured in the article. Pransky produces short, comedic videos in which she treats various robot patients on the proverbial couch. Pransky's work raises a series of questions, including: What might it mean to focus on robot anxieties about their interaction with, and use by, humans? Can a robot be anxious? To this end, in her debut video, "Robot in Crisis," she treats a robot with a tablet head who moves on two wheels.[2] He has just come from a bar mitzvah where he facilitated the telepresence in the US of grandparents in Israel (we learn that the grandparents are physically located 7,500 miles away from San Francisco, where the celebration was taking place). The robot laments, while sprawled on Pransky's office couch: "All the people! People everywhere! Pushing things at my screen! Laughing at me! . . . People speaking a foreign language took over. . . . [We see the robot dancing the hora.] I can't dance! I don't even have legs. I was scared." Pransky then assures the robot that he did well and enabled the humans to have a good time, even as she diagnoses him with multiple personality disorder and shifts the focus to his feelings.

The question of what might make robots anxious in their interactions with humans interestingly flips the problem of what makes humans anxious about techno-futures on its head. In Pransky's fantasy of the feeling robot, it is not humans who need to fear the robots, but robots who need to fear humans (or at least humans who put no thought into how their interaction with robots has effects in the world). According to Pransky, "'Robotic Psychiatry,' while touching upon the subjects of artificial intelligence (AI), human/robot interaction, etc., is still a humorous and entertaining medium intended to bring awareness, integration, and acceptance to a society in which there tends to be a constant undertone of disbelief or distrust when it comes to the topic of robotics."[3]

In mainstream robotics and AI, robot emotions are predominantly approached through the programming of predetermined drives that stand for emotions, with the aim of facilitating human interaction with robots. Pransky's fantastical formulation of robot anxiety as a product of interaction with humans stands in provocative relief to the reality of emotional programming in robotics. If robots can properly read and display certain emotions (happy, sad, angry, confused, etc.), thinking in the field goes, then the integration of robots into daily lives will be more easily accomplished and bring more pleasure to humans using robots in various social and occupational milieus.

The field of human–robot interactions (HRI) focuses on bridging the distance between the machine and the human by programming machines to display predetermined emotions that will be easily intelligible to human users. Yet, as we argue in this chapter, the racial-colonial logics undergirding the technoscientific conception of possible emotional drives (and their usefulness in human–robot interaction) reaffirms, rather than disrupts, the technoliberal capitalist logics of service and automation that uphold the supremacy of the liberal subject. This is achieved through the engineering imaginary of human sociality that determines the programming objectives of emotional states in the field of social robotics. Dwelling on the historical particularities of the development of social robots in the field of human–robot interactions enables us to assess the operations of the surrogate human effect in what Lucy Suchman has called artificial person-hood (or the engineering fantasies behind artificial personhood and how it is delimited).[4] As we propose, artificial personhood and the programming impetus to map the emotional drives attributed to human personhood are shaped by and contribute to the coloniality of the fabric of the social realm within technoliberalism. We further observe that social robots are designed as a mirror to prove to us that the apex of human evolution is our ability to perform the existence of an interior psyche to the social world. Social robots can only mirror human states. The position of the robot is always inferior because it must perform affect transparently, and therefore cannot perform proof of the hidden interior psyche that would indicate equality to the human. In this model of social robotics, human uniqueness is preserved, and machine sociality is positioned to enrich human social life. The imaginary of what a human is in social robotics could expand and redescribe the human and its social worlds. Instead, dominant thinking in the field has replicated a singular racial imaginary of what the human is and what it can aspire to be because of its basic Darwinian logic.

In our analysis, we closely read an array of roboticists' writings on social emotional robots, with a particular focus on what is considered the first sociable robot, Kismet. Though now retired from the MIT robotics lab, Kismet is the model for all subsequent social robot projects, like Jibo, which we discuss in chapter 3. Analyzing the engineering logics of Kismet allows us to understand how the imperial-racial legacies of a Darwinian emotion-evolution map are ingrained in the constitution of human–machine social-ity and the coconstitution of the machine and the human. The chapter opens with the history of how social roboticists themselves approach emotion and sociality. We examine how the engineering imaginary behind mainstream social robotics draws upon a normative human morphology in which a pre-given representation of the world is already subsumed. This representa-tion of the world assumes an ontological divide between an interior (self) and an exterior (environment) that mirrors the mind–body duality. Next, we excavate the racial underpinnings in the construction of emotional in-telligence in robots, turning to the importance of Darwin's work on emo-tions and evolution, and its racial presumptions, which persists as part of the normative morphology informing the contemporary engineering of emotional robots. Sociable robots enact the surrogate human effect as sur-rogate *affect* within technoliberalism, not in the sense that they are engi-neered to replace human–human interaction, but in the sense that they are engineered as mirrors to human interior states. The racial grammar of the surrogate human effect/affect, in this sense, starts with the engineering of the interiority/exteriority divide in conceptions of the self. We conclude the chapter with an exploration of the necessary ways we must expand the notion of sociality and personhood delimited by the racial-imperial frame of emotion in the field of robotics.

Emotional States and Artificial Personhood

In the 1970 article "The Uncanny Valley," Masahiro Mori made the case that "human beings themselves lie at the final goal of robotics."[5] Mori argues that up to a point, as robots appear more humanlike in movement and physical form, human responsiveness to and empathy for the machine increases. However, at a certain point, the positive responsiveness turns to revulsion (the valley). Mori uses the example of the prosthetic hand designed to look like it has veins, tendons, muscles, and fingerprints to describe this valley: "This kind of prosthetic hand is too real, and when we notice it is prosthetic, there is a sense of strangeness."[6] This is a "negative familiarity." Mori's thesis is that as robots are designed to look and move like human beings, the closer the resemblance, the higher the peaks and valleys of the uncanny (that is, the greater the resemblance to humanity, the greater the risk of a deep repulsion toward the machine on the part of the human). Thus, he urges designs that offer a sense of familiarity but that nonetheless preserve a perceptible difference from the human.

The ideal proximity–distance formula attempting to avoid the pitfalls of the uncanny valley introduced by Mori underpins the technoliberal frame undergirding the programming of emotions at the heart of social robotics. On the one hand, the field of social robotics emphasizes the goal of robot autonomy (as movement and emotional responsiveness to human use) that supports the fantasy of service to the human without human labor. Programming emotions into machines is about making machines just lifelike enough to be the most useful and pleasant for human use. On the other hand, the field of social robotics also works to erase human involvement in the production (and not just programming) of robot emotional responsiveness that reaffirms liberal notions of sociality as based on independent and disconnected rational actors in the world. The notion of robot autonomy in social robotics is based on continually reaffirming its difference from human autonomy. The robot ethicist Kate Darling has defined the social robot as "a physically embodied, autonomous agent that communicates and interacts with humans on a social level."[7] As she explains, "autonomy in robotics can be understood to mean as little as the ability to perform tasks without continuous human input or control."[8] Cynthia Breazeal, the MIT roboticist who helped pioneer the field of social robotics, has argued that sociable humanoid robots fundamentally shift the conception of machine autonomy.[9] Breazeal notes that

> traditionally autonomous robots are designed to operate as remotely as possible from humans, often performing tasks in hazardous and hostile environments (such as sweeping minefields, inspecting oil wells, or exploring other planets). Other applications such as delivering hospital meals, mowing lawns, and vacuuming floors bring autonomous robots into environments shared with people, but human–robot interaction in these tasks is still minimal.[10]

In contrast to these more common uses for robots, which treat them as tools, the goal of Breazeal's research since she was a graduate student working with Rodney Brooks at MIT has been to rethink robots as human partners. This partnership is about the production of robot persons who complement (without disrupting) notions of human personhood.

The human–robot complementarity envisioned within social robotics reaffirms a developmental hierarchy that positions the human at the summit of an evolutionary gradation of emotional development. Programming emotions into robots is a final frontier in robotics because emotion is increasingly viewed as a sign of intelligence more complex than that displayed by most computers and commonly used robots in industry. Reasoning and logic are displaced by "affective competencies" as the hallmark of the human, and thus also of what will be lucrative in the field of robotics.[11] Programming emotion into robots in a way that will mirror normative emotional states in humans depends on the physical presence of robots in a world shared with human beings. For Breazeal, as the leading social roboticist in the US, the physicality of the robot—in terms of both its morphology and its material presence in space—is crucial to constructing a social relation between human and machine. Breazeal writes that "facial expression, body posture, gesture, gaze direction, and voice" all work together to produce sociable robots, making them suited to communication with humans.[12] Moreover, unlike disembodied systems, which interact

Figure 4.1. Rodney Brooks and Cog, Brooks's humanoid robot.

with humans through speech or text entered by humans, embodied systems with which a human interacts, such as a robot or an avatar, must
sense the human. The perception of the human for social robotics involves
more than sensing the physical presence of an object, as, for example, Rodney Brooks's robots that focused on mobility without cognition were programmed to do. Rather, the problem of perception also involves sensing
social cues and responding to human emotional states (see figure 4.1).[13]

That social robots are humanoid in their morphology does not mean
that their emphasis is on replicating the human form or function exactly.
As Breazeal, Brooks, and other roboticists working on embodied robots
designed to interact intimately with humans have stressed, the *"magic"* of
robots is that they are *not* humans, and the engineers' goal is not to replicate what humans already do well.[14] For instance, Kerstin Dautenhahn,
who works on robots designed to interact with children with autism in
the AURORA project, treats autonomous mobile robots as potentially thera-
peutic precisely because they are not human. She argues that humanoid
robot platforms can help children who can't handle the unpredictability of
human behavior.[15] She thus posits that "using humans as models for creating believable technology . . . is not a universal solution for designing robotic social actors."[16] For Dautenhahn, the lifelike agent hypothesis, which
suggests that artificial social agents interacting with humans must imitate

or mimic human forms and actions, unnaturally restricts robotic function.[17] To illustrate this point, she utilizes an example contrasting a stick and a plastic sword, positing that most people can imagine doing more kinds of things with a stick than a plastic replica of Excalibur.[18] Humanlike robots, similar to the plastic sword, Dautenhahn explains, limit the scope of users' possible interpretation of how to interact with them. The anthropomorphic form restricts expectations about robot behavior and cognitive complexities.[19] In this sense, the notion of the social robot as a partner to the human that is distinctively not human echoes J. C. R. Licklider's well-known conception of "man–computer symbiosis" formulated in 1960.[20]

At the same time, there is something about the human form and the human tendency to anthropomorphize robots that has led engineers to conceive of new possibilities for human–machine sociality when machines take on humanoid form. In much of his early work on mobile robots, Brooks designed machines that had an insect-like form. As he explains it, "there turned out to be two good arguments for building the robot with human form. The first is an argument about how our human form gives us the experiences on which our representations of the world are based, and the second argument is about the ways in which people will interact with a robot with human form."[21] While Brooks cautions that "the robots we build are not people," for him, the main impetus toward the humanoid form in social robotics is that "we [humans] have been programmed through evolution to interact with each other in ways that are independent of our particular local culture."[22] These "programs" range from making and averting eye contact to how close we should stand to another person. According to Breazeal, machines must mirror biological systems in order to seem plausible to a human.[23] As she contends, social robots learn by imitating a human user. In a similar line of argument promoting anthropomorphism in specific contexts, Kate Darling argues that the human tendency to anthropomorphize robots more than they do other kinds of objects (such as toasters) is useful insofar as people are more likely to accept and integrate new technology in their lives when they can relate to it on a human level.[24]

To achieve relatability, Breazeal's understanding of artificial intelligence includes rather than excludes emotional intelligence (being able to read other people's internal states and to respond in ways established to be socially appropriate). For her, intelligence is also something that develops over time through learning systems rather than something that can be fully preprogrammed. Breazeal's Kismet, the prototype of her Sociable

Figure 4.2. Kismet's emotional expressions.

Machines Project, is organized around the processors, input devices, and hardware required to animate Kismet's "face."[25] These programmed drives, modeled on Darwinian biological drives, determine the robot's performance of an emotional state: "The signaling of emotion communicates the creature's evaluative reaction to a stimulus event (or act) and thus narrows the possible range of behavioral intentions that are likely to be inferred by observers"[26] (see figure 4.2). Kismet's sensory and perceptive abilities were engineered to be concerned with interior rather than exterior states (expression rather than action) in both human and machine. In comparison to screen-based interfaces or even nonhumanlike robots, anthropomorphic embodied systems like the robot Kismet that interact with humans have the benefit of paralinguistic communication, including gesture, facial expression, intonation, gaze direction, or body posture.[27] However, to engage human interaction in the first place, experiments show that facial expression is best combined with postural evidence of attention.[28] A physical body that can manage both of these things therefore aids in certain forms of engagement and interaction.

Kismet, with its stuffed animal-like big eyes and bat-like ears, was built in the late 1990s, but is to this day probably the best-known sociable robot because of how much publicity it has received over the years. From PBS documentaries to articles in the *New York Times*, what appears most fascinating about Kismet is its apparent ability to respond to people's emotions, tone of voice, and facial expressions.[29] In publicity videos from the 2000s, Kismet always seemed to know when it was being scolded, or when to smile back at a visitor to the MIT lab. Most importantly, Kismet was programmed to be sociable, with "the same basic motivations as a 6-month-old child: the drive for novelty, the drive for social interaction and the drive for periodic rest. The behaviors to achieve these goals, like the ability to look for brightly colored objects or to recognize the human face, were also part of Kismet's innate program. So were the facial behaviors that reflected Kismet's mood states—aroused, bored or neutral—which changed according to whether the robot's basic drives were being satisfied."[30] Breazeal's team describes the hardware that enabled the robot to communicate its emotional states based on the satisfaction of its drives as follows: the robot was designed

> with perceptual and motor modalities tailored to natural human communication channels. To facilitate a natural infant–caretaker interaction, the robot is equipped with visual, auditory, and proprioceptive sensory inputs. The motor outputs include vocalizations, facial expressions, and motor capabilities to adjust the gaze direction of the eyes and the orientation of the head. . . . These motor systems serve to steer the visual and auditory sensors to the source of the stimulus and can also be used to display communicative cues.[31]

Kismet was also designed to learn like a child. To achieve this goal, Kismet's cuteness (the large eyes, protruding ears, and fuzzy eyebrows) was meant to encourage users to treat it as they would a baby. In turn, inspired by theories of infant social development, Breazeal sought to make Kismet

> enter into natural and intuitive social interaction with a human and to eventually learn from them, reminiscent of parent–infant ex-

changes. To do this, Kismet perceives a variety of natural social cues from visual and auditory channels, and delivers social signals to the human through gaze direction, facial expression, body posture, and vocal babbles. . . . These capabilities are evaluated with respect to the ability of naïve subjects [non-engineers unfamiliar with Kismet's software and hardware] to read and interpret the robot's social cues, the human's willingness to provide scaffolding to facilitate the robot's learning, and how this produces a rich, flexible, dynamic interaction that is physical, affective, social, and affords a rich opportunity for learning.[32]

While it would be easy to dismiss the sensing-action approach to artificial intelligence as associating physical abilities with childlike intelligence and higher cognitive skills (like playing chess) with adult capabilities, it is worth noting that at present it is the more basic robot abilities (motion, facial expression, and human interaction) that are considered much harder to engineer and program than those that could be viewed as "higher abilities," or skills difficult for adult humans to master.

At the same time, Kismet's sociability and the effectiveness of its affective display are based on the appropriateness and legibility of the facial expressions exhibited by both robot and human. The programming of emotion is thus about the correct translation of interior states to an exterior social world. Put otherwise, the goals of hardware and software alike are sensing and action that reaffirm already given conceptions of a world shared between robots and humans. According to Lucy Suchman, "the circumstances under which objects are apprehended as persons" are important to consider in relation to the broader problem of how personhood is figured in the Euro-American imaginary.[33] Suchman argues that autonomy and rational agency as the two key attributes of personhood are culturally specific to post-Enlightenment Europe. For instance, writing about Brooks's and Breazeal's approaches to mobility and space, as well as their emphasis on the physical presence of the social robot in the field, Suchman points out that ideas about "the world" and the bodies that move around in that world are pre-given (thus "the world" is seen as an unchanging, preexisting entity, as is the body and what is taken as its recognizable human form). Suchman further notes that the discreteness of bodies and world carries forward into the programming of emotions in social robots.

When taken as "discrete states," emotions become available for analysis and replication.[34] Overviewing the prehistory of social robotics in medicine, Suchman points out that

In the laboratory, the drive to produce clear, compelling representations of emotional states (as measured through various physiological changes), led to the coconfiguring of imaging technologies and subjects. "Good and reliable subjects" were chosen for their ability to display clearly recognizable emotions on demand, while those that failed to produce unambiguous and therefore easily classifiable behaviors were left out of the protocol. . . . These technologies produced a catalogue of emotional types, normalized across the circumstances of their occurrence (e.g. as anger, fear, excitement), and treated as internally homogeneous, if variable in their quantity or intensity. Inevitably, standard readings developed based on experimenters' prior experience and cumulative data. And as inevitably, particularly in the context of the early twentieth century when these experiments flourished, categories of emotion were mapped to categories of person, affording (often invidious) comparison across, for example, men on the one hand, "women and Negroes" on the other. . . . At the same time, this was an economy that circulated through, but was discursively separable from, specific bodies.[35]

In this history, we see not only how emotions render the body itself as a machine, but also how notions of autonomy and discreteness limit what it means to figure a feeling artificial person.

As part of the project of upholding the separateness and boundedness of categories of vitality and action (including those of human and robot, emotion and affect, and bodies and worlds), social robotics as a field has gone to great lengths to erase human labor in representations of affective computing. Suchman notes that in numerous promotional videos of the work done, for instance, at the MIT robotics lab,

the efficacies demonstrated are narrated as portents of developing capacities, from which the rest of human capabilities will logically and inevitably follow. Together these rhetorical leaps conjure into existence an imaginative landscape increasingly populated by "socially intelligent" artifacts, approaching closer and closer approximation to things that both think and feel like you and me. Through these modes of erasure of human labors and nonhuman alignments, the autonomous artifact is brought into being.[36]

The archive of robot evolution and development is thus an archive of ever-increasing levels of humanness that machines must demonstrate precisely by stripping the human from the scene of its very production. The stripping of human–machine intra-activity preserves a bounded concept of the human not changed through its interaction with machines, and vice versa. Moreover, as Suchman insists, "The 'humanness' assumed in discussions of the potential success (or the inevitable failure) of attempts to replicate the human mechanically is typically a humanness stripped of its contingency, locatedness, historicity and particular embodiments. But if we take 'the human' to be inseparable from specifically situated social and material relations, the question shifts from 'will we be replicated?' to something more like 'in what sociomaterial arrangements are we differentially implicated, and with what political and economic consequences?'"[37] Suchman is here calling for a relational approach to personhood. As she concludes, "The point in the end is not to assign agency either to persons or to things, but to identify the materialization of subjects, objects and the relations between them as an effect, more and less durable and contestable, of ongoing socio-material practices."[38]

In light of this, we might ask: How do notions of evolution and emotional states, which "translate" interior states to the exterior world through embodied expressiveness, inform the design of mechanical psychic states? In what ways does the programming of emotional responsiveness replicate the racial, gendered, and colonial logics of the surrogate effect within technoliberalism? Why must we program the servile other to be emotionally expressive, and how does this programming reaffirm the centrality of the liberal subject while masking the work that goes into its ongoing reaffirmation?

Race, Evolution, and the Space of Affect in Social Robotics

For social roboticists like Cynthia Breazeal, programming emotion is about mapping and putting on display what she has called the "affective space" within machines. As we propose in this section, the affective space is the space within which the racial grammar of the surrogate effect enters and structures the technoliberal imaginary of the social in social robotics. The programming of emotion is based on a utilitarian model of relation between man and machine that is about making the machine fully transparent and legible for predictable, easy, and pleasurable use within capitalist relations. Because Kismet is all head (without a body) and its sole purpose is to elicit and display emotion (internal states), as we argue, it is an exemplary robot through which to address how what we are calling the surrogate effect is engineered into ideas about robot futures.

Cynthia Breazeal has noted that the three fields most pertinent to the development of sociable robots are psychology, ethology (the study of animal and human behavior from a biological perspective), and evolution.[39] To produce a mechanical creature that was dynamic and capable of emotional and social learning, Breazeal found inspiration in "ethological views of the analogous process in animals" and, in particular with the programming of Kismet's drives and emotional states, she sought to mimic biology in her development of artificial emotional intelligence.[40] Because of this, Kismet was self-consciously engineered to reflect the history of human evolution. As Breazeal explains, theories of evolution that derive from Darwin's writing on emotion were crucial for her programming and design. Darwin's 1872 theory "that a few select emotions are *basic* or *primary*" thus centrally informs the Kismet project's version of artificial emotional intelligence necessary for meaningful human–machine interaction. According to Darwin, "emotive signaling functions were selected for during the course of evolution because of their communicative efficacy. For members of a social species, the outcome of a particular act usually depends partly on the reactions of the significant others in the encounter."[41] Breazeal takes this theory up in her work, asserting that the primary emotions "are endowed by evolution because of their proven ability to facilitate adaptive responses to the vast array of demands and opportunities a creature faces in its daily life. The emotions of anger, disgust, fear, joy, sorrow, and surprise are often supported as being basic from evolutionary, developmental, and cross-cultural studies."[42]

We can read Breazeal's approach to emotional expression in robots as a kind of social Darwinism with a twist that characterizes the surrogate effect of sociable (obeyant) robots. In social robotics, evolution is equated with the capacity for affective communication, a capacity that undergirds the possibility of human–machine interaction within capitalism. The problem is that this imaginary of the social, while it stands poised to redescribe the human and its social worlds, replicates a singular social-scientific imaginary of what the human is because of its basis in Darwinian logic. As Henri Bergson writes, "the line of evolution that ends in man is not the only one. On other paths, divergent from it, other forms of consciousness have been developed, which have not been able to free themselves from external constraints or to regain control over themselves, as the human intellect has done, but which, none the less, also express something that is immanent and essential in the evolutionary movement."[43] Theories of knowledge and theories of life are inseparable, because "a theory of life that is not accompanied by a criticism of knowledge is obliged to accept . . . the concepts which the understanding puts at its disposal."[44]

We suggest that the epistemological-ontological concepts left unquestioned in the opportunity to reimagine social interactivity between man and machine offered by social robotics are that of the racial and that of the social. Racial demarcations of bodily difference undergird the teleology of the post-Enlightenment human that ends in Europe, and thus reaffirm well-established social norms and orientations. What Breazeal terms the "affect space"[45] of the robot, that is, how the robot's valence, stance, and arousal dictate the robot's pose to indicate an emotional state to its human users, is mapped and imagined through this familiar history of human evolution—one that is based upon a racial mapping of the globe meant to decode universal human emotion. We can thus think of social robotics, with Kismet as its prototype, as being haunted by what Banu Subramaniam has evocatively called Darwin's ghost in the modern biological sciences, which we here extend to the engineering world.[46] As Subramaniam writes, "scientific theories of variation come from contestations of very particular variation in humans—of sex, gender, race, class, sexuality, and nation—variation that was understood in a political syntax of its times. These histories have shaped and been shaped by science and are indeed constitutive of science."[47] Crucially, "delving into these histories reveals the profound debates around eugenics, about desirable and undesirable bodies—those doomed to sterilization, enslavement, or colonization or deemed perverse, deviant, pathological, or deficient."[48] This is significant, because biology, along with the modern sciences more broadly, is "doomed to play out its eugenic script" unless it openly comes to terms with this history.[49] Thus, it is important to unearth what is unspoken about evolution in the programming of Kismet, as this matters a great deal for the kinds of scripts that are embedded in notions of the "social" of the social robot.

Of course, "affect space," because it is a space interior to a body, at first appears to be quite different from the imperial space of racial command and control through which human bodies were differentially classified and categorized in evolutionary biology. At the same time, because decisions about the "appropriateness" of Kismet's emotive responses is based on ideas of "social interaction as originally argued by Darwin," it is important to explore in greater detail how Darwin's initial arguments about emotion emerged out of a colonial racial-spatial mapping of the universal human.[50] In *The Expression of the Emotions in Man and Animals* (1872), Darwin's concern was with the question of whether the same expressions and gestures prevail across "all the races of mankind," especially "with those who have associated little with Europeans."[51] Just as Kismet's emotional expressions are paralinguistic and meant to be "cross-cultural," so too was Darwin interested in the innate and instrumental expressions that translated across what he considered to be distinct human races. To accomplish this, his method was to obtain answers to a questionnaire from "missionaries or protectors of the aborigines" in Australia, which he considered to be "several of the most distinct and savage races of man."[52] Other groups observed included the Maoris of New Zealand, Malays, and Chinese immigrants in the Malay archipelago, as well as Indians in Bombay.[53] Darwin emphasizes that "negro slaves" in America were not observed because of their close contact with white men, but some of the "wildest tribes in the West of the New World" were included in the data. In addition to ensuring that the races observed were as distant as possible from Europe so as to be certain of the universality of particular emotional expressions, Darwin sought to account for instances, as in India, where the respondents had "much difficulty in arriving at any safe conclusions, owing to [the natives'] habitual concealment of all emotions in the presence of Europeans."[54] Other races were seen to transparently reveal their emotions. The more transparent the display of emotion, the more distance was mapped between that emotive state and the civilizing control over affective display. Thus, in addition to "primitive" races, Darwin also observed infants, the insane, and animals—those "untouched" by civilization.

To be clear, Darwin was not arguing that the display of emotion by the insane, the infant, or the savage were the same as that of European man, who stands at the apex of the evolutionary scale. Rather, it is that the perceived unrestrained, if not base, display of the basic and instinctual emotions revealed something about the primal instincts that made European man, even if these base emotions were now subject to civilizational taboos. Thus, the most primitive, savage, or irrational beings, through their spatial-geographic as well as temporal distance from European man, are able to prove the universal. In short, exteriority (geography, race, and distance from Europe) determines the way in which the interiority and psychic state are displayed and made legible. For example, Darwin describes the unrestrained weeping of the Maori who are provoked "by very slight causes," and he retells a story of a New Zealand chief, who ostensibly "cried like a child" because European sailors had spoiled his favorite cloak by soiling it with flour.[55] Similarly, the insane and children "give way to their emotions with little restraint."[56] However, as Darwin explains, frequent efforts to restrain weeping can check the habit.[57] Due to such cultural limits, what is "natural" is not as transparent in Englishmen, who, according to Darwin, rarely cry except from the "acutest grief."[58] Similarly, whereas children all around the world pout, adult Englishmen do not pout, while adults in New Zealand are perpetually pouty and sulky, as characterized by the protrusion of the lower lip.[59] For Darwin, those who are "deprived of reason," like animals, the insane, and those races distant from European man, become "brute in character" as they are reduced to base emotion.[60] Yet these primordial habits are nonetheless the basis of our evolution as a species, and European man must recognize these emotions as still a part of himself in spite of civilizational checks. Darwin thus insists that all races are one species, as evidenced by "the close similarity in body structure and mental disposition of all races of mankind."[61]

The seeming contradiction in Darwin's theory of emotions, namely that it seeks to establish the universality of certain emotions through a global mapping of racial distance from the space of civilization and reason, produces the racial grammar of Kismet's affect space and therefore the imaginary of human–machine relationality at the heart of "the social" in social robotics. We do not intend here to simply point out how the racialization of childhood in Darwin is replicated in the ideal of childlike learning in Kismet and other robots utilizing learning systems, nor do we mean to imply that Kismet is itself somehow racialized in its expressiveness. Rather, we are interested here in the epistemological/ontological racial structures that persist, and endure, in the imaginary of human–machine sociality exemplified by social robots.

Darwin's writing on emotion and evolution, the basis for Kismet's expressive response to human interaction, exemplifies the representational moment of modern scientific knowledge about Man, in which, as Denise da Silva argues, the transparent "I" of post-Enlightenment modernity is consolidated through and against the affectable "I"—the non-European mind.[62] Da Silva writes that scientific signification, through the colonizing act of simultaneously naming difference and marking for elimination that which is named, has deployed the racial to produce seemingly transparent modern subjects. She describes this transparency thesis in two parts: first, the taxonomic impulse of science as "engulfment" is a subsumption into the modern through the violence of naming; second, and simultaneously, as preexisting lifeworlds enter into the field of modern scientific representation, they become marked for "murder, total annihilation." Modern scientific knowledge thus institutes "the necessary but haunting relationship between an I instituted by the desire for transparency (self-determination) and the affectable, always already vanishing others of Europe that the scientific cataloguing of minds institutes."[63]

Affectability, which in Da Silva's sense is "the condition of being subject to both natural (in the scientific and lay sense) conditions and to others' power," is in Darwin's theory of emotion quite literally on display in the non-Europeans' unconditioned show of emotion. They are in a state of being easily moved by minor events (like having flour poured over their cloaks) because they lack civilizational checks. Through Darwin's documentation and scientific capture of this unrestrained emotion in Europe's others, European Man emerges as a universal (transparent) subject. Writing about nineteenth-century US projects of social engineering, Kyla Schuller has argued that "impressibility," or "the capacity of a substance to receive impressions from external objects that thereby change its characteristics," came to signal an evolutionary accomplishment. The habits of civilization over time were understood to result in a refined nervous system that marks the transformation of "animal substrate into the cultural grounds of self-constitution."[64] According to Schuller, sentimentalism, in addition to impressibility, demonstrated that civilized bodies were able to respond to their cultural and physical environment through disciplining their "sensory susceptibility." This evolutionary accomplishment marked a form of life that needed protection from the threat of bodies more primitive, as marked by impulsiveness and low sensitivity. These qualities of the so-called primitive body performed evidence of its incapability of evolutionary change and its imminent obsolescence.[65] Thus, race and sex differentiated people's capacity for impressibility such that the self-possessed liberal subject came to be defined through its capacity for change (as opposed to reactivity).

In Kismet's interaction with humans, Kismet is the emotional object because it is moved (affected) by human users, yet at the same time it can only react to the human—it is incapable of itself changing (or evolving) over time—an attribute tethered to the fully autonomous human subject. It works to read what human users want and strives to make human users *not* work to read its emotional states. For instance, Breazeal argues that for the robot to emerge as an enjoyable and interesting social partner for the human, it must be *transparent* to the human user.[66] As Breazeal explains, for Kismet to be successful, it had to demonstrate that it was learning and responding appropriately through emotive expressions that were legible as properly those of fear, anger, happiness, surprise, and so on to the human user, as well as to make its intents clear to its human interlocutors. At the same time, the robot had to be endowed with a "socio-cognitive architecture" capable of inferring mental states of the users, including beliefs, intents, and desires.[67] In the case of sociable robots, the ultimate goal is predictive "mind-reading," by which engineers mean that a robot can employ "multiple world and agent models," including estimates of what a human partner can see from their vantage point.[68] Through its display of emotion as direct response, the machine thus makes itself transparent as responsive to the human, while cameras and algorithms are used to make human intent, desire, and belief legible to the machine so that it can respond appropriately.

The racial legacies of this human–object relation are evidenced not just in the fact that, following Darwin, emotional transparency is a racial state, but also in the sense that Kismet's emotions are subjugated to human intent. In a different context, Sara Ahmed has argued that "'being emotional' comes to be seen as a characteristic of some bodies and not others, in the first place. . . . We need to consider [instead] how emotions operate to 'make' and 'shape' bodies as forms of action, which also involve orientations towards others."[69] Ahmed argues for a model of a "sociality of emotion" that questions the presumption of emotion as based on interiority.[70] This means not just that "emotions create the very effect of the surface and boundaries that allow us to distinguish an inside and outside in the first place," but also that "emotions are crucial to the very constitution of the psychic and the social as objects."[71] Treating the social as an object with potentially fluid boundaries, Ahmed's point could allow for a rethinking of how Kismet and its human interlocutors can remake borders through human–machine interaction. However, Breazeal's conceptualization of emotions as biologically pre-given states that can be programmed into objects (robots) fails to do so.

This is in part because the project of social robotics, as Breazeal's post-MIT entrepreneurial endeavors demonstrate, is about the proliferation of technoliberal capitalism. In this sense, the programming of emotions into robots is about an economic logic. Neo-Darwinian instrumentalization of evolutionary models lends itself perfectly to an economic calculus. As Carla Hustak and Natasha Myers argue in the context of Darwin's writing on orchids and this text's reinterpretation by neo-Darwinians, "adaptations are made legible as rational choices in an economic logic that aims to maximize fitness advantages while calculating the lowest energy expenditures. These logics have become hegemonic in the contemporary life sciences."[72] They explain that "evolutionists tend to fetishize economic logics, random mutations driving generational change, and functionalist accounts of adaptation."[73] These economic logics translate seamlessly into the logics programming emotions as drives. They describe an informatics model of communication in which binary questions are the only pathway to produce information—for example, "self or non-self" or "male or female." In this way, any stream of information can be mapped as a timeline of consecutive and always binary choices, and the "amount" of information can always be quantified in bits. They suggests that "[t]his informatic model of communication, which so readily fits into the calculating logics of a militarized economy, may itself be read as a defensive strategy that protects researchers from breaching the constrained forms of agency permissible inside neo-Darwinian models of evolution."[74] The unexpected transformations of the human in its interaction with machines, as well as unforeseen functions and uses of social robots, are thus excluded in the reaffirmation of the racial logics of the Darwinian evolutionary model utilized by Breazeal and others to formulate the "social" of "social robotics."

The import of neo-Darwinian logics for the study of emotion, and the racial implications of these logics as they structure the social, extend well beyond the field of robotics. They are ingrained in the coloniality of US global dominance. In a *New York Times* opinion piece about the 2003 photographs from inside Abu Ghraib, one of which featured the white, female soldier Sabrina Harman smiling and giving a thumbs up next to the brutalized body of a dead prisoner, film director Errol Morris interviews the behavioralist Paul Ekman, who, like Breazeal, uses Darwin's theories of emotive display to interpret expressions and emotions that may not be obvious.[75] In the interview, Ekman explains that based on the breadth of the smile, the creases around the eyes, and so on, we can understand that Harman's smile is posed—as one does for a camera—and not a genuine expression of pleasure. Ekman insists that what is horrifying about the image of Harman next to the dead body of an Iraqi prisoner is that the spectators of the image want to smile back. Thus, rather than asking who killed the prisoner, we ask, why is Harman smiling? In this association of white womanhood with innocence, those horrified by the image of Harman's smile are implicated because they cannot distinguish a genuine smile from a posed one. Ekman asserts that

> The most important thing in terms of adaptation is for you to know that the other person is either actually or simulating enjoyment. And that was more important than whether they really were enjoying themselves. The fossil record doesn't tell us much about social life. All one can do is to say there is no really good facial signal that evolved. Now when people laugh in a phony way, that's a little easier to pick up. But even then, most of us want to hear good news. We don't want to hear bad news. So we're tuned to it. We're very attracted to smiles. They're very salient. But telling the feigned from the genuine, we're not good at that for any emotion, for anger, fear. It takes quite a lot to train a professional, a National Security or law enforcement professional (we do quite a bit of that) to be able to distinguish between the two. There are no clear-cut obvious signs. So what must have been important was to know what a person was intending, not what they were feeling.[76]

Ekman's neo-Darwinian position on the ability to read genuine versus simulated emotions (and to therefore inscribe intent) reasserts agency in the realm of whiteness by privileging interiority (intent) over exteriority (display). The correspondence of display and true emotion is rendered complex when tethered to white bodies (especially those working in the service of US empire building). In contrast, the correspondence between interior states and emotional display is understood to be exact only when apprehended on bodies of not-quite or not-yet human others (those whose interiority is not yet fully developed).

The surrogate effect of programmed emotions, engineered into robots' affect space, assert transparency and affectability as the condition of obeyant, racialized sociality within technoliberal capitalism. The definition of "the social" built into the robot's interactive design holds the affectable "I" as the condition of possibility for the transparent "I." Breazeal's emphasis on robots as human *complements* that are not the self-same as the human carries forward the foundational need for the not-quite-human affectable "I" as the mirror through which the transparent "I" can understand what it means to be fully human. Affectability is also the mark of subservience. As Breazeal herself put it, robots are a mirror to reflect what it means to be human, including what our values are, precisely because they are the not-quite-human other in our imagination.[77] Thus, when Elizabeth Wilson makes the point that the coassembly of machines and affect is not a thing of the future, but that it is "the foundation of the artificial sciences," we might also observe that this engineering of emotion into the artificial person is at the same time a reassertion of the racial in the figurations of the mechanical/artificial nonhuman and human boundary.[78]

Wilson is critical of the Kismet project. She explains that in Breazeal's Kismet, emotional states in the other are deduced intellectually rather than systematically known, and "mutuality is executed rather than sensed."[79] She demonstrates that the commitment to the view of the human mind itself as a binary system (the neuron is either firing or it is not) continues to the present, even in projects like Kismet that make claims to emotion and sensing/perception rather than the intellect.[80] Wilson connects this to her treatment of Alan Turing's earlier and foundational commitment to intellect as being directed through fundamentally logical "channels of symbolic language."[81] She concludes that Breazeal, like Turing, is drawn to emotion but loses her nerve in the end.

Though Wilson's critique of the theory of emotion behind the design of Kismet as not *truly* expressing emotion, but rather intelligence as programmed drives, opens up a space, as does Ahmed's writing, to rethink the relationship between emotions and the social by disregarding the centrality of the Darwinian evolutionary model in roboticists' imaginaries of sociable, emotional robots, Wilson also misses the role of the racial in that design. By neglecting the racial ontoepistemological field that governs understandings of both exterior environments and interior states, as well as both thinking and feeling, Wilson inadvertently replicates the erasure of race as the condition of possibility for articulating and imagining affective states as interiority. Most significantly, Wilson focuses in on the exclusion of one and only one of the eight basic affects named by the psychologist Silvan Tompkins: shame. All of the other affects named by Tomkins, including interest, enjoyment, surprise, anger, fear, contempt, and distress, follow Kismet's six primary emotions and three drives.[82] For Wilson, the exclusion of shame, which provides important conditions for socialization, reduces Kismet to "a robot designed to regulate the production of shame and dampen down its transmission."[83] This makes Kismet "drive-dependent [and] cognitively oriented," and with its "shameless affect," close to the "classical AI paradigm" that erases the question "Can a robot feel?" in favor of the question "Can a robot think?"

Given our argument that highlighting the importance of the Darwinian model is crucial to understanding *both* the paradigms of thought and affect in artificial worlds and their relationship to human worlds, perhaps Wilson's dismissal of Kismet as an older model is too simple. When he discusses shame in *The Expression of the Emotions in Man and Animals*, Darwin is limited in his abstraction of the emotion when he poses the question of whether shame always forces a blush where the color of the skin allows it. Crucially, the display of shame (even if shame is felt) is disallowed to those darker-skinned humans because it cannot be perceived by another. This is significant given Darwin's focus on the *expression* of the emotion, rather than the internal states, in the function of emotion for evolution.

Toward a Feminist, Anti-Racist Theory of the Social

Walter Pitts, one of the central characters in Elizabeth Wilson's study of the homosocial early history of AI science, with its masculine platonic romances and cults of personality, was committed to arriving at a mathematical theory of human brain function.[84] This was thwarted by what turned out to be a pivotal experiment in frog vision, specifically on the neurophysiology of how a frog sees. The authors of the study, including Pitts, discovered that the cells in the frog's eye were actively perceiving the world: "The eye speaks to the brain in a language already highly organized and interpreted, instead of transmitting some more or less accurate copy of the distribution of light at the receptors."[85] This discovery went against the binary hypothesis of how the brain processes sensory input, in which both the brain and the subsidiary mechanism behind visual perception were digital—on/off. In addition to being a turning point in the history of cybernetics, in which the boundary of the human becomes a central concern,[86] the frog study undermines the ability to presume a digital model of the mind, and it undermines a separation between the materiality of the body and the perception of the mind. Even so, the digital model of intelligence, and the predominance of the mind, and of thinking, as the domain of human social interaction, has prevailed even in emotional robotics.

The history of social robotics is a productive place to think about how social relations enacted through normative bodies and models of perception and communication can carry the politics of political modernity into technological modernity. The field's normalization of embodiment, the centrality of "thinking," the privileged sensorium, and the physical performance of attention and obeyance are being mainstreamed as a gendered and sexualized racial formation in the arena of domesticated social relations.

While human neurophysiology has adapted its understanding of affectability to include numerous platforms, we have shown how social robotics has stayed close to a model that insists on an interior "affect space," an imagined three-dimensional interior space in which emotional processes can be precisely located, the contents of which are communicated socially through exterior expression. We have argued through a review of seminal writing on the engineering of emotion in robots that the affective "space" of the robot is therefore where the surrogate effect, as a racial grammar, comes in to organize the technoliberal imaginary of the social in social robotics.

The role of embodiment in perception and cognition of the sensate world has been a central stake in the work of feminist theory and queer theory that protests the givenness of gendered and sexual social relations, and so we bring this protest to the reliance in social robotics on anthropomorphic embodiment, and its entrenching of associated norms of gendered and racialized obeyance. Importantly for imagining the terrain of affect and emotional robotics, the field of phenomenology, beginning with philosopher Merleau-Ponty, has rejected Descartes's subject–object divide to support what we would argue is a materialist and feminist critique of the exclusion of embodied experience as a fundamental platform of sociality. Merleau-Ponty argues that individual perception is socially entangled with our empathetic and relational understanding of the embodied experience of an other, and it is through enacting embodiment that we understand and inhabit our own bodies.[87] He argues that the bodies of others are intelligible through one's own bodily experience and skills: "I can mimic the gestures of another, for example, without thinking about our respective positions in objective space. I immediately sense my own movements as equivalent to the movements of the other, and I experience my own body as an open system of an infinite number of equivalent positions in the orientation of the other."[88] He is not arguing that it is the mind's mediation that makes the relation possible, but rather that through embodiment we must perceive ourselves through the other in a way that is inherently social.

Merleau-Ponty's observation helps us question why how to program "thinking" into social robotics has remained a central concern. Despite the frog-eye experiment and its revolutionary impact on the understanding of the human neurophysiology of perception, in emotional robotics it is still assumed that the true content of communication is held in the mind, and that the body signals this message through a set of common codes. This "digital" or informatics and binary mode of communication and perception, where the message is either correctly conveyed or fails to be conveyed, insists on the mind as mediator, separate from the idiosyncratic body. Social science research on emotion continues to privilege an assumed universal platform of biologized pathways, predicated on universalized drives, that were the basis of programming Kismet and later social robots' reading of human cues and their performance of nonverbal communication with humans. Without a model that privileges thinking, the transparency of the (white, Euro-American, able-bodied, civilized) human subject cannot be so easily assumed through the model of interiority.

If we reject the exteriority/interiority divide (as well as the boundedness of bodies and worlds, humans and machines), we might come to observe that humanity, as an essence, is always surrogate, not an essential quality of interiority. This means that humanity can only exist in a relation, so that the surrogate effect remains suspended "between" the nodes of a relation or multiple relations. Influenced by feminist STS scholarship, we therefore insist that it is only through reading this inherently social process of "making surrogate" that we find something like "humanity" to begin with. For example, in *When Species Meet*, Donna Haraway's feminist materialist nonhuman exceptionalist reformulation of coproduction, the "encounter" is asserted as constitutive of who and what is encountered. In other words, she argues that all relations are materially and semiotically constitutive of that which is related. As the book's title implies, Haraway is interested in the meeting between individuals and groups of "species of all kinds, living and not."[89] She describes the importance of the encounter as shaping both subject and object as "a continuous ontological choreography."[90] The concept of "encounter value" thus elucidates the process of creation in the relation.[91]

What is clear in the juxtaposition of the racial colonial history of the science of emotion with the programming of emotion into social robots is that both enable the imaginary of a universalized "platform" of sensibility and drives common to the human, despite shifts in neurophysiology and interventions from phenomenology and feminist and queer theory. We must recuperate the desire to reimagine the human, and even the social, through interactions with artificial worlds, even as we consider how our imaginaries, ingrained as they are in post-Enlightenment epistemes, often fall short in ways that have a significant effect on human populations that continue to be seen and treated as not yet fully human.

To emphasize the racial scaffolding of both "thinking" and "feeling" as conduits of the forms of sociality that designate humanness is not to replicate a paranoid reading of AI. Rather, it is to emphasize the importance of keeping ongoing colonial legacies in mind when analyzing the engineering imaginaries of nonhuman worlds. A feminist decolonial and anti-racist approach to the role of emotions and affect in the social needs to challenge this ongoing history. As Ahmed argues, the field of emotion creates the effect of interior and exterior, as well as the very concepts of "the psychic" and "the social" as dependent on the effect of interior and exterior, respectively.[92] Disturbing these connections, then, must become part of a redefinition of the social that is expansive enough to account for how socially relating *itself*, as demonstrated in this chapter through the interiority/exteriority paradigm, can contain the effect/affect of making others the surrogate for the transparently human subject.

As an example, we turn to the work of artist and engineer Kelly Dobson, whose project Omo works to expose, and to explode, the limited understanding of emotion and sociality we have critiqued in this chapter. Omo disrupts the celebration of productivity and functionality at the heart of ideas about human–robot interaction by defamiliarizing the fantasy of human autonomy and difference. One of several machines she designed as part of a project titled "Machine Therapy," Omo is a relational object conceived as a critique of companion robots. Dobson argues that carebots are modeled to represent a whole creature or person, one that often vacillates between slave and authority. Companion robots are generally designed to be soothing, medicalizing, and normalizing. They are intended to help a person meet a narrow range of medically defined statistics of health and wellness. Dobson criticizes carebot design in that bots are not meant to present their own anxieties and neuroses, but are intended simply to comfort, serve, survey, and medicate. Dobson's response to these problems is Omo. She writes, "Instead of troping whole anthropomorphic others such as nurses, grandchildren and pets, Omo is designed to be experienced by a person as in between being part of the person—something akin to a prosthetic or body part—*and* a separate thing."[93] Omo operates by expanding and contracting at different volumes and rates of repetition in reaction to the pressure stimulus of a breathing or otherwise vibrating body it contacts.

Dobson describes Omo as a transitional object, saying, "The machine may sometimes, effortlessly and smoothly, seem a part of the person holding it, and at other times the person may perceive the very same machine as a separate object, as its own being."[94] Omo may seem in some situations or at some times to behave like an organism, creature, friend, or pet, but this is an unstable relation built into the object's design. Omo is also not designed to always be soothing and subservient. Instead of an emphasis on a singular and correct display of emotions, it has "errors" and noise built into its programming that lead it to act erratically and break down. Omo can both mimic a pressure-breathing pattern and lead the human user in breathing. If the person stays entrained with Omo, the object may influence the user's breathing pattern and thereby his or her physical and emotional state. Due to noise in its programming, Omo is not predictable and is not designed only to soothe.

Dobson's critique is important because human functional equivalents, like sociable robots, both reflect and shape fantasies of human obeyance. As people interact with and reflect on these caring machines and soothing machines that defamiliarize the form and function of care as they are tethered to capitalist propagation, people are, in effect, being trained into a new understanding of caring and soothing. They thus challenge the utilitarian and neo-Darwinian logics of emotional drives, imagining objects as being outside of a relation of service to the human.

5. Machine Autonomy and the Unmanned Spacetime of Technoliberal Warfare

The Mine Kafon, created by Afghan designer Massoud Hasssani and de-signed in Kabul, is a wind-powered metal ball with bamboo legs. Afghan-istan is estimated to contain nearly 10 million active land mines. The Kafon prototype is devised to be heavy, so that as the wind moves it across war-ravaged lands laden with mines that have not yet been activated, the sweeper can trip the mines with its weight as it tumbles. The bamboo spikes are designed to blow off the porcupine-like ball upon impact with a land mine, even as its center remains intact. Each Kafon can handle up to four blasts. It is also affordable—the cost to replace each unit is under forty dol-lars. Hassani's goal is to clear the world of land mines within a decade with low-cost tools such as this wind-powered minesweeper. The Kafon, which moves without human intervention or control, through wind power, is in a sense autonomous. Yet, although it moves autonomously, Hassani's design is not a robot, nor is it a high-cost or high-tech design. Moreover, unlike many other kinds of military technologies, the Kafon can be read as a car-ing technology, performing the feminized work of cleaning up the ravaged landscapes produced by war, rather than taking part in the act of killing.

Because the minesweeper also cares for human life, it untethers caring from the purview of the human actor. As a technology, the Kafon thus unexpectedly brings technologies of care to the field of war, and, wind-powered as it is, we see it as critically positioned in relation to post-Enlightenment concepts of autonomy.

We open this chapter with the example of a caring, autonomous military technology to apprehend the particularities of the racialized and gendered dynamics of the category of autonomy at the heart of technoliberalism. Within the Enlightenment paradigm, technology has been understood as a tool for implementing social and economic development, and as an extension of the human. In both these senses, technology reaffirms human autonomy and agency—one is fully human when one has the power to make decisions about how technology is used or made productive. Put otherwise, the liberal human subject uses technology to be an agent of historical, economic, and social progress.

Given the entanglement of the liberal humanist production of human autonomy through the conceptualization of technology as a tool and extension of the human body that is nonetheless completely separate from the boundedness of that body, it is unsurprising that specters of technologies that could act autonomously—that is, without human oversight or decision making—have been perceived by tech industry leaders and antiwar activists as the ultimate threat in artificial intelligence and learning systems development. Autonomous lethal weapons, or "killer robots," which could make decisions about who is a potential target and when to strike, loom large on the horizon of present-day modes of waging war. Yet, when it comes to machine autonomy, affect, or care, is something that is understood to be the purview of the human. After all, autonomous technologies still cannot *feel*. At the same time, care work and affective labor are feminized and devalued forms of human work. It is also work that has historically been invisibilized as work, upholding the myth of the autonomous human.[1] This is why the seemingly incongruous pairing of machine autonomy in the military field and care embodied by the Kafon pushes at the bounds of post-Enlightenment figurations of the fully human.

In this chapter, we analyze military technologies as technologies that most clearly render machine autonomy as threatening in US public debates about the dangers that technology poses to the human as species. Post-Enlightenment notions of autonomy produce and uphold the liberal subject through a series of erasures (including, as we've discussed in prior chapters, erasing work that is deemed dull, dirty, repetitive, or uncreative). Such erasures also work in the present through the prevalence of the fantasy of "unmanned" military technologies, such as drones and killer robots. By addressing drones (semiautonomous weapons) and so-called killer robots (or fully autonomous lethal weapons), we problematize the notion of autonomy, arguing instead that we can see military technologies as "cobots," or collaborative robots, that continually remake the bounds of the human. As we propose, the myth of the "unmanned" and its production of human autonomy is not only based on the racialized dynamics of servitude and the fear of rebellion (or the other's coming into consciousness as a historical subject), but also on fantasies of remote control that obscure present-day modes of occupying space and time as the imperial techniques of technoliberalism. These modes of occupation based on the surrogate effect of military technologies are underwritten and justified by the obscuring of human labor in the myth of the unmanned. The threat of machine autonomy thus foregrounds the need to attend to race and gender as the post-Enlightenment paradigms through which the process of coming to consciousness as an autonomous liberal subject takes shape.

As we elaborate in this chapter, when a robot "evolves" from an enslaved object to a potential killer (meaning that it can turn on the maker or master, Man), the robot can be recognized as having gained autonomy or consciousness even as it becomes a dangerous object that must be stopped. At the same time, the fear conjured by the potential for violence enacted by a nonhuman actor toward a human elicits a desire for the domestication and subsumption of that other back into subjection and servitude. Thus, as we propose, killer robots and robot slaves are linked in a racialized dynamic. Autonomy, in the sense of coming into a historically specific agential subject position, is a racial fetish of post-Enlightenment thinking, as decolonial thinkers like Sylvia Winter, Frantz Fanon, and Gayatri Spivak among others have asserted. Our theorization locates servitude and slavery as central to the grammars of autonomy in both its post-Enlightenment and cybernetic formulations. The autonomous subject who is free to act upon the world (and other bodies and objects in that world) has historically been defined against the enslaved. It is thus not surprising that the killer robot and robot slave are positioned along a racialized spectrum of autonomy that connects the post-Enlightenment figure of the autonomous human and more recent engineering-based logics of technologies that enact the surrogate human effect in the field of war.

Autonomous and semiautonomous military technologies reveal how contemporary anxieties around nonhuman or machine autonomy continue to conceive of autonomy in relation to the racialized inheritances defining objecthood, property, and self-possession. Even as machine autonomy would appear to disrupt human supremacy, in fact, it reiterates the racial scaffolding upon which autonomy for the post-Enlightenment human subject was consolidated. At the same time, although human operators control drone trajectories and weapons discharge, as an unmanned aerial vehicle (UAV) (an aircraft with no pilot on board), the drone still participates in the ongoing creation of the mythology of the "unmanned" that has characterized the recent coproduction of the human with surrogate human technologies. UAVs, or drones, with US-based remote operators, have been part of standard US military operations in Iraq and Afghanistan since 2002.[2] The common understanding that there are no humans involved in drone warfare was recently denounced publicly by former US drone technician and whistleblower Cian Westmoreland, who said, "Although often portrayed as unmanned, drones are in fact 'hypermanned,' . . . with thousands of people from different agencies and different countries feeding into a global network."[3] The bracketing of human participation to produce the surrogate effect of "unmanned" UAVs, and the invisibilization of the networks required to operate drones, have both been articulated as central to the military culture of dehumanization necessary to materialize "the target." For drone operators, coming to consciousness requires a confrontation with this culture even as it represents a failure to perform the task of remote-control or human-free killing efficiently (that is, without the bad feelings that come from the recognition of the humanity of the target).

By foregrounding post-Enlightenment conceptions of autonomy that ground the liberal subject, this chapter speculates about what it might mean to consider the racialized dynamics of military "cobots." This is not simply an assertion about human–machine co-assemblage. Rather it is an argument that the particular configuration of autonomy and military technologies, through the construction of killable populations or targets, builds on post-Enlightenment imperial tools of spatial and temporal command and control. This perpetuation and expansion of imperial command and control occurs even as the "new" technoliberal imperial form proclaims its innocence through the twin conceptions of human autonomy and the unmanned in contemporary warfare. Positing that both autonomous and semiautonomous weapons are cobots disrupts notions of autonomy that at once produce myths of unmanned warfare and the racialized objecthood tethered to servitude and affective labor within technoliberalism. To push at the bounds of the surrogate human effect in the field of war that undergirds post-Enlightenment notions of autonomy, we thus attend to care as a limit point of technoliberal forms of violence. First, we analyze how it is the *impossibility* of care attributed to machines that demarcate the fully human in debates about machine autonomy. Next, we turn to reconfigurations of the spatial and temporal sensorium as a part of an emergent technoliberal imperium, and finally, to the designation of drone operators who care about their targets as themselves failing at their jobs because that care disrupts the efficiency attributed to automated warfare. This dual failure of care in the sphere of war sets up our discussion of the myth of the unmanned as a technoliberal postracial update to present-day imperial conquest.

The Racialized Dynamics of Machine Autonomy

Rather than representing a break with the Enlightenment notions of human autonomy in its apparent separation from subjectivity, machine autonomy inherits the racial imperial legacies of earlier philosophical formulations. Though seemingly freed of human bodies, machine autonomy is delimited by racial epistemologies—they are the invisible foundation of the software and hardware that materialize the killer robot. Like the drone, the killer robot activates a social imaginary of human-free assassination, in which the killer robot spares soldiers' lives to eliminate "targets" that have already been evacuated of humanity. This fantasy produces a notion of "appropriate" killing that is in service of a hierarchical chain of command. At the same time, unlike the drone, which is only semiautonomous, the killer robot as a fully autonomous weapon, one that makes decisions about targets without human oversight or command beyond its initial programming, contains a threat of rebellion against its makers. The potential for full autonomy and fully automated warfare thus exposes the contradictions structuring the ideals of human-free killing. First, killer robots both trouble and reaffirm the conceptual boundedness of human autonomy consolidated through the history of racial slavery and the masters' fears of slave rebellion, as we argue in this section. Second, the fantasy of human-free killing in contemporary remotely waged wars, which simultaneously reaffirms the humanity of those who have command and control over space and designates killable "targets" within that space, is the operation of the surrogate human effect.

Cybernetics first defined machine systems autonomy as the ability to demonstrate purposeful and goal-oriented behavior. The study of communication and automatic control systems in machines and living things/systems, cybernetics arose in concert with computer science in the late 1940s. Lucy Suchman and Jutta Weber note that "while the Enlightenment concept of autonomy is grounded in the idea of a free and self-aware subject, one which can self-determinedly and consciously choose its maxims, the cyberneticians explain purposeful behaviour not in rational-cognitivist terms but as a pragmatic physiological mechanism that can be automated: 'A torpedo with a target-seeking mechanism is an example.'"[4] The goal is pre-given, but the system decides how to pursue it. Suchman and Weber conclude that this version of autonomy is "grounded in a profound semantic shift" from that of the Enlightenment because it is "defined as the capability to explore random real-world environments, by which sometimes unforeseen and useful behaviour might emerge."[5] Put otherwise, behavior-based robotics claims to build autonomous robots because these can handle unpredictable situations in real-world environments, not because they are free or have self-awareness. *This is an autonomy without consciousness.*

The challenges that machine autonomy pose to the post-Enlightenment tethering of the liberal subject's consciousness as a historical actor and that subject's unique ability and freedom to act upon the world are not without contradictions. For instance, Suchman and Weber propose that with "autonomous weapons systems it becomes even more obvious how autonomy is configured as a self-sufficient, adaptive and self-determined performance on the one hand, and preprogrammed, fully automated execution under perfect human control on the other."[6] Suchman contests both of these notions of autonomy, citing the STS scholarship on material agency, assemblage, and actor network analysis: "In the words of feminist theorist Karen Barad, 'agency is not an attribute but the ongoing reconfigurings of the world.'"[7] The main question we ask about machines, as Suchman and Weber conclude, should thus not be focused on automation versus autonomy, but rather on "what new forms of agency are enabled by contemporary configurations of war fighting, and with what political, ethical, moral, and legal consequences."[8] Yet, while these new forms of agency hold the potential to lead "to a reconceptualization of autonomy and responsibility as always enacted within, rather than as separable from, particular human–machine configurations," this reconfiguring is at best an incomplete and failed displacement of Enlightenment racialized epistemes of the human.[9]

The racial inheritances in the concept of autonomy are reflected in the engineering goals of machine autonomy, subsuming the threat of nonhuman autonomy to the supremacy of the human. They do so by figuring killer robots as enslavable precisely because machines, in their inhumanity, threaten to be more rational, powerful actors than the original rational autonomous subject, European man. Robots' enslavability thus reiterates the human right to command objects as tools of imperial expansion. As material labor in service to the human, machine autonomy must be domesticated and objectified. The incompleteness of machine (nonhuman) autonomy as a break from the Enlightenment version of autonomy surfaces at this intersection of conceptions of autonomy as the physical command over space on one hand, and as material labor on the other.

We can think of the domestication of the killer robot as a reaffirmation of human command in the face of the threat of machine autonomy that endangers the supremacy of the liberal in the world of objects and those who have been objectified. That is to say, the surrogate human effect of autonomous technologies can be celebrated as long as objects remain obeyant to the human. They can neither achieve consciousness nor disrupt the prescribed set of relations describing appropriate uses for machines in distinct contexts (such as domestic work or the work of killing). Examples abound, as military robotics and service robotics have been interconnected in recent engineering ventures, thus delimiting engineering imaginaries that value technologies based on their surrogate human effect, linking to servitude. iRobot, the company that created the most popular service robot currently in use worldwide, the Roomba automated vacuum cleaner, is also the producer of PackBot. PackBot is "one of the first widely deployed military robots. Light enough to be carried by a single soldier and controlled with a handset developed directly from game system controllers, the PackBot is used for reconnaissance bomb disposal, currently in service in Afghanistan and Iraq."[10] The connection between the vacuum cleaner and the military robot that stands in for a broader link between the servant robot and the killer robot is part of engineering imaginaries that inform the design of both kinds of machines to require the development of rule-learning systems (including social learning in interactions with humans) and similar kinds of mobility (picking up and manipulating a variety of objects). The interconnection of the engineering of care and service work, on the one hand, and violence, on the other, follows from the Enlightenment logics of enslavability and freedom that continue to bookend conceptions of autonomy in racialized and gendered terms.

The military is currently funding a number of research projects to produce autonomous robots in order to diminish the need for human soldiers to risk their lives. These projects demonstrate the racialized and gendered terms of service that are subsumed into technoliberal fantasies of human autonomy and command. For instance, Boston Dynamics, one firm that has received US government funding for developing military robotics, gained notoriety in October 2013 when it released a video of a six-foot-tall, 320-pound humanoid robot named Atlas running freely in the woods (figure 5.1). Atlas is a prototype first introduced in the media as standing in for soldiers in the theater of war, and more recently touted as a rescue robot, intended to replace firefighters, police officers, and agents cleaning up nuclear waste.[11] In response to the idea that Atlas was designed as a first responder for various crises, such as nuclear disasters, numerous media reports about the robot insisted that it was much easier to imagine this huge machine as a super-soldier than as a humanitarian actor. One reporter proclaimed, "At 6-foot-2 and 330 pounds, the U.S. Army's new Atlas robot just might give Arnold Schwarzenegger a run for his money. . . . Atlas can turn knobs, manipulate tools, and yes, fire machine guns. Last month, developers showed off its remarkable dexterity by having it mimic the workout routine made famous in *The Karate Kid*."[12]

The Defense Advanced Research Projects Agency (DARPA) funds the Atlas project, like other Boston Dynamics robot development projects. Yet, in spite of the fact that killer robots appear to be an unprecedented innovation, as one magazine reported, Atlas is not unprecedented, but is an outgrowth of already existing trends in the military.

About 40 percent of the U.S. aerial fleet consists of unmanned combat drones, and the Air Force now trains more drone operators than bomber or fighter pilots. Robotic ground vehicles have also flooded the battlefield, with more than 6,000 deployed in Iraq and Afghanistan to haul gear, climb over obstacles, and provide advanced

Figure 5.1. Atlas tested in the woods.

reconnaissance. The military is now testing a more advanced version called a Legged Squad Support Systems robot that looks like a mechanical mule, and can carry 400 pounds of cargo over uneven terrain. With even more futuristic devices in the pipeline, some officials are estimating that up to 25 percent of infantry forces could be replaced by robots in the coming decades. Like aerial drones, most of the current systems are operated by remote control, but experts say some devices will soon be capable of carrying out designated tasks with minimal human oversight.[13]

Atlas is a platform that has the potential for autonomy, as do its predecessors, because of its programming through so-called rule-learning systems. These are the basis for the software for self-exploration and error correction, based in randomized planning algorithms that allow a robot to figure out the best way to grasp or move an object, and imitation learning, whereby a robot can watch how a human person handles an object or how an animal moves.[14] The idea is that the robot can get immediate feedback on the efficacy of an act in relation to the programmed goal. It can record "rules" of engagement to use in future actions, thereby learning through trial and error.

Because of its potential for autonomous movement and decision making, Atlas has spurred a variety of both utopic and dystopic fantasies about robots that don't require human oversight to kill. These fantasies offer insight into the ways in which enslavability and freedom figure human autonomy as a racial project that is continually threatened when tools seen as external to man and made solely for his use threaten to rebel. Soon after Boston Dynamics introduced Atlas to the public in 2013, the robot was quickly shown to be capable of being programmed to clean up messes with large brooms and vacuum cleaners, spotlighting how it can "use tools designed for human use."[15] Atlas accomplishes pressing the vacuum pedal or getting the empty bottle into the waste because of military robotics goal-oriented programming, where details are left to machine trial and error. Yet the engineering parlance of "degrees of freedom" used to describe Atlas's range of motion and mobility in fact show Atlas in a relation of unfreedom (that is, in service to) the human.

The choice of housework as a way to showcase the programming and mobility of Atlas's learning systems is not innocent. Angela Davis has written about gendered structures of housework, explaining that "Invisible, repetitive, exhausting, unproductive, uncreative—these are the adjectives which most perfectly capture the nature of housework."[16] Although housework and its devaluing as the "invisible" and the "unproductive" is a historical outgrowth of capitalism, as Davis argues, "prevailing social attitudes continue to associate the eternal female condition with images of brooms and dustpans, mops and pails, aprons and stoves, pots and pans."[17] The gendered nature of Atlas's vacuum cleaner demos defangs the killer robot by asserting the feminine range of its motions that supplement the masculine ones suitable to the field of war. Debbie Douglas, the director of MIT's museum that features a permanent exhibit of that institution's key contributions to the field of robotics, mentioned in an interview with us that "Robots need good PR [public relations] so that people will not fear them."[18] The video of Atlas sweeping and vacuuming, when juxtaposed to the video of Atlas holding a machine gun or practicing martial arts to demonstrate its capabilities as a replacement for human soldiers on the battlefield, domesticates the killer robot so that it is not feared by humans. As one fantasy mashup photo of Atlas and Rosie the robot maid of the future from the *Jetsons* cartoon (figure 5.2) signposts, Atlas may even have a place in the home and family.

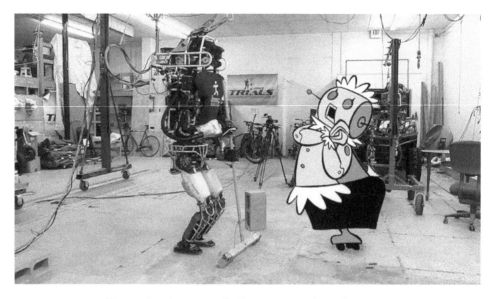

Figure 5.2. A domesticated Atlas meets Rosie from *The Jetsons*.

At the same time, the ostensibly seamless transition from automated housecleaning to automation in global contact zones defined by perpetual warfare exposes the accretion of race and gender within service labor. Lisa Lowe points out that what we know as "race" and "gender" are the traces of an intentional process of forgetting the violent conditions of indenture and enslavement that made the particular freedom of the modern liberal subject possible.[19] Atlas's facile makeover from killer robot to housecleaner reinvigorates the definition of violence as the racial and hetero-patriarchal potency for enslavability even as Atlas's "PR" erases the gendered and racial-historical forms delimiting its engineered capacities toward autonomy. Atlas's housecleaning is meant to demonstrate that machine autonomy can never surpass the autonomy of the post-Enlightenment human. Rather, machine autonomy can only be "gifted" as permission to the machine, so long as the permission for autonomy is subsumed within and reaffirms the supremacy of Enlightenment ingenuity and the autonomy of the liberal human subject. Whereas earlier imperial civilizing projects claimed to develop the potential for full autonomy (as self-rule or participation in the capitalist global economy) through colonization, the technoliberal imaginary now seeks to gift autonomy to the nonhuman machine. In both instances, the secondhand autonomy, or the autonomy that is permissible along the sliding scale of humanity, is enacted through a sovereign violence that distinguishes itself from the kind of violence that the killer robot could itself wield.

Atlas's autonomy means that the threat of rebellion is always just under the surface in any demonstration of physical prowess and sovereign movement. To perform the limits to machine autonomy, Atlas's mobility and agility are tried and tested through pushing, prodding, and hitting that put it in its place as unfree object of human control. Here is how one journalist described a video demonstrating Atlas's agility:

> The [Boston Dynamics promotional] video . . . [shows] Atlas bending down to pick up 10-pound (4.5 kilograms) boxes and pivoting its torso to place each package on a shelf. In another instance, a human handler uses a hockey stick to push Atlas off balance. The robot stumbles backwards (but catches itself) before regaining its balance. Next, an employee pushes Atlas down from behind. The curled-up robot (lying flat on its robotic face) is able to push itself up—first to its "hands" and "knees," before righting its torso and then pushing up on its feet—all without help from a human or a tether.[20]

Atlas is not the only robot to undergo pushes, shoves, and even kicks. In fact, all of the Boston Dynamics robots are kicked to demonstrate their agility and mobility. While these interruptions are meant to demonstrate the robots' capacity to right themselves, these moments also provoke visceral reactions among viewers of the video—the motion of the robot's feet, in both their technical maneuvering and timing, have contradictory connotations. On the one hand, this movement inspires fear of the object rising up against its maker. On the other hand, it inspires awe as the object offers a glimpse of itself as vital and animate. That all of the Boston Dynamics robots must undergo a kicking references back (in spite of explicit erasures and decontextualization) to the fact that robots are designed to be the property of, and in service to, the human. As Despina Kakoudaki has explained, robots are historically a fantasy for an ongoing system of enslaved labor.[21] Yet it is in the specter of (and in some sense, the desire for) the rebellion of killer robots that the threatening, and then reaffirming, racialized structures of autonomy and consciousness reside.

Both the scale of civilizations under European territorial colonialism and the projected timeline of AI and machine autonomy occupy a temporality of the "not yet." Like the notion of human development projected onto that scale of civilization and humanity, where the colonizer assumes the task of cultivating the colonized to further develop its own humanity toward the end goal of European civilization, the imaginaries of robot rebellion reference the racialized inheritances structuring a possible interiority or self-consciousness that has yet to fill a robot's metallic shell.[22] These racialized inheritances have to do with attributing interiority and feeling through an imaginary of likeness—in the case of Atlas, its anthropomorphic shell. The "not yet" measures proximity to the human as a course of progress toward what is already recognized by the inheritor of that colonizing subject position, the (white) Euro-American liberal subject, as a course of progress toward the conclusion of the human, and of the realization of the consciousness that denotes entry into historical progress.

Given the cultural field of robot revenge imaginaries, in which killer robots are predicted by the rebellion of enslaved and violated or mistreated machines, it is unsurprising that a series of spoof videos followed from the Boston Dynamics official YouTube postings. One video, titled "I Heart Box," uses the Boston Dynamic demo video but features a voiceover narrative in which Atlas professes its love for the boxes it lifts. As the engineer hits and pushes away the box as the beloved object, Atlas attempts to regain it. Finally, Atlas gives up on regaining the box and is shown walking out the door (intended by the original Boston Dynamics video to showcase how well Atlas can turn a door knob—a task notoriously difficult for robots to do well). With the voiceover, this departure is given new meaning. Atlas mumbles that it loves the box, just as the engineer loves his daughter. The last words in the video are those of the engineer swearing at Atlas and demanding to know what the robot meant (though of course, there is the implication the daughter could be harmed).[23] In another video dubbed the "Atlas Swearing Module" Atlas shows up to work in a foul mood and refuses to put up with the engineer's torment (figure 5.3).

In this context, the robot Atlas rising to its feet after being kicked, though of course programed by human engineers, startlingly conjures robot free will and consciousness—this is the act of a machine standing up to a human master. This movement is made all the more frightening by Atlas's origins as a weapon. In commentaries accompanying this video, viewers joked, "Have fun with that hockey stick for now, meatbag, in 10 years YOU'LL be the one being toyed with!" and "Well at least we know who the robot is gonna murder first when it gains self-awareness."[24] Others, in a more serious tone, wrote, "Maybe someday the robot will upload and process this, and process [that] I am not your slave."[25]

Figure 5.3. "Atlas Swearing Module."

The sense of "justice" in machine liberation from a preconscious (enslaved) state to a state of consciousness and emancipation, even as such liberation threatens to unleash violence back upon the human, can nonetheless be read as a reaffirmation of the liberal subject rather than its unmaking. This is because the human is still the only point of reference for an ethical relation with the nonhuman or not-yet human. Given our argument that robots, even potentially life-threatening killer robots, engender the surrogate human effect as the racial form of servitude delimiting the engineering imaginary of machine autonomy, we might consider how these revenge fantasies about Atlas rising are based on an object's proximity to the human. According to one article, feelings are more readily attributed to robots when they look lifelike:

> Researchers have found that when people watch a robot being harmed or snuggled they react in a similar way to those actions being done to a flesh-and-blood human. In one study, participants said they felt negative emotions when they watched a human hit or drop a small dinosaur robot, and their skin conductance also showed they were distressed at the "bot abuse." When volunteers watched a robot being hugged their brain activity was the same as when they watched human–human affection; even so, brain activity was stronger for

human–human abuse versus human–robot violence. "We think that, in general, the robot stimuli elicit the same emotional processing as the human stimuli," said Astrid Rosenthal-von der Pütten of the University of Duisburg Essen in Germany, who led that study. The research was presented in 2013 at the International Communication Association Conference in London.[26]

Empathy and relatability tie the register of "humanness" to the colonial scale of civilization, which ranked the humanness of racialized people across the globe by their perceived proximity to European lifeworlds. Fantasies of robot rebellion following a coming into consciousness thus reiterate a Hegelian notion of freedom and self-actualization in this new technoliberal theater where a twenty-first-century master–slave confrontation takes place. If robots can prove that they are like "us," they then enter into the realm of universality as self-possessed subjects, and in their potential humanity can be recognized.

The Racial-Imperial Spacetime of Drone Warfare

Whereas machine autonomy reaffirms the autonomous figure of the human as the ethical agent of historical progress, tethering servitude and violence to the nonhuman while vesting the potential for empathy and care with the human, drones erase human action and accountability from the fantasy of technoliberal warfare. Drones must accomplish this erasure even though, in reality, they are more akin to cobots because they are only semiautonomous, keeping the human in the decision-making loop over life-and-death decisions in the field of war. The coloniality of technoliberal warfare, made manifest as drone warfare, depends upon the myth of "unmanned" violence. Technoliberal warfare is founded upon, even as it refigures, both settler colonial and territorial racial logics of human-free lands available for capitalist development. For example, in European territorial colonialism, the command of space was secured through the installation of governance and control by resident colonial rulers and their armies. In settler colonial formations, the permanent residence of the colonizer is one of the primary technologies of gaining command of space, and the protection of that residence becomes the grounds for punitive violence that enables expansion of the physical and political settlement. Following these logics, the present-day exercise of remote command through roboticized warfare materializes the "target" as something distinct from a human "enemy combatant."

The fantasy of remote control manifest in the drone takes the human colonist/interventioneer out of the frame of action. Put otherwise, the surrogate effect of drone warfare frames command over space as being human-free. In this sense, remote control is not only a project of saving the lives of would-be conquering soldiers (while rendering the lives of those to be conquered even more dispensable than in earlier conquests). In addition, the drone mediates the relation to humanity of both potential killers (soldiers) and potential targets (victims or casualties). The surrogate effect of roboticized warfare thus renders the fantasy of remote control as a reconceptualization of empire that, by being human-free, can disassociate its power from earlier modes of colonial conquest.

Remote control fantasies tethered to drone warfare seek to "relieve humankind of all perilous occupations. Miners, firefighters, and those working on the atom, in space, or in the oceans could all be converted into remote-control operators. The sacrifice of live bodies [is] no longer necessary. Once living bodies and operative ones [are] dissociated, only the latter, entirely mechanized and dispensable, would now come into contact with danger."[27] Drone-enabled remote killing can initially be justified as a humanizing violence that will end terror because of the distance and dissociation it enacts between the operators and targets. Technological asymmetries racialize so-called danger zones that must be commanded remotely so that they are figured as salvageable for the spread of liberal rule and freedom. The reconceptualization of spatial command through remote control is thus part and parcel of the advent of what Atanasoski has elsewhere theorized as a humanitarian imperialism that emerged with US Cold War liberalism.[28] As we argue here, it has been further developed in the age of the "war on terror" as technoliberal warfare whose humanization takes the human out of the frame of imperial command and control.

Remote control, which removes human accountability while reifying human autonomy (command), is the surrogate effect of technoliberal warfare. The history of engineering remote control over distant spaces involved the reconceptualization of space as a field for (non)human actors. According to Gregoire Chamayou, the modern history scaffolding the proliferation of drones and killer robots can be traced back to the engineer John W. Clark, who theorized "telechiric machines" as a "technology of manipulation at a distance."[29] For Clark, "In the telechiric system, the machine may be thought of as an alter ego for the man who operates it. In effect, his consciousness is transferred to an invulnerable mechanical body with which he is able to manipulate tools or equipment almost as though he were holding them in his own hands."[30] Yet, as Chamayou points out, what is missing in this second body is "the living flesh of the first body," as that vulnerable first body is "removed from the hostile environment."[31] The surrogate effect of the machine thus fundamentally shifts the space constituted by warfare:

> Space is divided into two: a hostile area and a safe one. . . . The hostile zone . . . remains a space that is left derelict but which, as a potentially threatening area, definitely needs to be kept under surveillance. It may even be exploited for its resources, but it is not, strictly speaking, to be occupied. One intervenes there and patrols it, but there is no suggestion of going to live there except to carve out new secured zones, bases, or platforms in accordance with a general topographical schema and for reasons of security.[32]

Numerous military robots already in use reiterate colonial racial fantasies as surrogates for human on-the-ground presence. These robots are a complement to the distancing dynamics of the drone's aerial presence. In 1947, the US Army and Navy pooled resources to conduct atomic tests at Bikini Atoll in the Pacific, and at that time announced plans for robotic aircraft, that is, drones, to "dive into the atomic blast to gather scientific data."[33] As early as 1964, NASA began plans for a "robot laboratory" to land on Mars—a precursor to the Mars rovers.[34] That the Mars rovers and drones continue to occupy our imagination of remote control over space today, and the fact that both still conjure various futures (one utopic, the other dystopic), speaks to the extent to which Cold War geopolitics still influence technoliberal conceptions of command and control. Now as then, engineering is positioned to solve the problem of how to command a space that it is physically (or temporally) impossible for humans to penetrate.

The surrogate effect of the machine that can move in so-called hostile environments (whether those of outer space or those in the field of war) is doubly racialized in function and form. First, the right of the liberal subject to those spaces previously impenetrable is yet again reasserted and reassured through technology. Technological progress, connected as it is as tool and mode of innovation to liberal notions of historical progress, affirms imperial expansion as the logic of engineering and use of machines meant for remote control. Second, the racial other, figured as compliant and kill-able, was the historical model for how drones were engineered to enact the surrogate effect, rendering them at once useful and disposable as tools. According to Katherine Chandler, the World War II navy drone engineers used Japanese kamikaze suicide bombers as the inspiration for imaginaries of drones as "American kamikazes."[35] As Chandler demonstrates, engineers relied on racial stereotypes of the Japanese to imagine remote control. At-tributes like unflinching obedience to a higher cause attributed to the Japanese became the basis for framing "radio control technology . . . en-visioned as more-than-human, unfaltering and compliant in its approach to death."[36] The racial other, like the drone, cannot possess autonomy and is thus reduced to object status (to be used in the field of war). The lack of autonomy as a characteristic of racial otherness was thus appropriated and built into the technoliberal infrastructures of command from a distance.

At the same time, to foster such engineering innovations, new spaces to command must constantly be conjured (not just for the sake of discovery, but for advancement, progress, and productivity). In addition to its promo-tion of Atlas, in the last decade the firm Boston Dynamics has showcased numerous videos of its prototyped military robots, all of which feature ter-rains hostile to human mobility as a way of demonstrating the physical prowess and agility of these new machines. In fact, the function of the majority of Boston Dynamics bots is described by how they move through space. The short overview of one of the company's smaller military robots, the Sand Flea robot (figure 5.4), is typical:

> An 11 pound robot that drives like an RC car on flat terrain, but can jump 30 ft into the air to overcome obstacles. That is high enough to jump over a compound wall, onto the roof of a house, up a set of stairs or into a second story window. The robot uses gyro stabilization to stay level during flight, to provide a clear view from the onboard camera, and to ensure a smooth landing. Sand Flea can jump about 25 times on one charge. Boston Dynamics is developing Sand Flea with funding from the US Army's Rapid Equipping Force (REF).[37]

Remarkably, though the company does not acknowledge this history, the name *Sand Flea* refers back to a military operation that was part of a US strategy for maintaining its imperial presence in Latin America. The training exercise preceding the 1989 US invasion of Panama was dubbed Operation Sand Flea. The title of this military operation designates the importance

Figure 5.4. The jumping Sand Flea bot.

of troop movement during the practice assaults—the "freedom of move-ment drills"—meant to assert the right of the US military to be present and engage the Panama Defense Force. These were highly conspicuous displays of US military presence staged to showcase the US prerogative to defend its rights to the Panama Canal. The seemingly innumerable drills and US movements were also designed to overwhelm the Panamanians in their ability to understand the US tactics in preparation for the US inva-sion to come.[38] The connection between mobility and imperial command over space is thus built into the engineering premise of advanced robotics. In the words of Boston Dynamics, robots must be about "mobility, agility, dexterity and speed" that exceed human ability.

In spite of the significance of machine mobility as the human-free com-mand over space, one of the major failures of robotics to date has been the slow pace of developments in robot dexterity (especially in comparison with the leaps and bounds of AI development in the last several decades). As a March 2017 article in *Wired* magazine wryly suggested, "If you're ever feeling down, do yourself a favor and watch some footage from the 2015 Darpa Robotics Challenge."[39] This challenge involved numerous scientific teams from various universities, all of which were supposed to program and modify the Atlas bipedal robot platform to maneuver through an obstacle course. The challenge was to make the robot successfully perform simple tasks, such as turning doorknobs and walking across uneven terrain. In contrast to the celebratory promotional videos released by Boston Dynamics, it became clear that the Atlas robot could not perform even the most humble tasks without difficulty. Thus, as the *Wired* article explains, "Our face-planting future robotic overlords could stand some improvements."[40] The article frames these improvements as evolutionary—taking movement to where the human form cannot go. In describing the company's newest robot, Handle, which moves around on two wheels and has two upper appendages (resembling a stallion when it lifts itself up on its hind legs), the article notes that

> What Boston Dynamics has done with Handle is take what natural selection has crafted—the human form—and turned it into a more efficient chimera. It's created an evolutionary marvel. . . . What's remarkable about Handle is that it has essentially one-upped evolution. Natural selection never saw fit to give a mammal wheels, for obvious reasons—you'd need a motor and bearings and a perfectly flat world and let's face it I don't need to tell you why natural selection never invented the wheel. But wheels are extremely energy-efficient. So much so that Boston Dynamics claims Handle can travel 15 miles on a charge. Just imagine a bipedal robot trying to stumble that far.[41]

On the one hand, Handle is here portrayed as breaking the pattern where the human form limits engineering and design imaginaries within robotics. On the other hand, Handle's function is still very much limited to its surrogate effect of affirming humanity as a project of evolving toward a technoliberal future where the human is removed from the degraded arenas of manual labor and killing, and instead nonhuman others populate warehouses and the field of war (two arenas that the *Wired* piece highlights as being improved by the advent of a robotic platform like Handle). In this sense, separating robotic evolution from human evolution reiterates the racial grammar of the surrogate effect within technoliberalism, as human evolution is an evolution away from the racialized (dirty, dull, disgusting, and dangerous) functions of service and imperial killing.

Moreover, disarticulating the technologies of remote control from human command over space refuses to view drones and military machines as cobots, thus reaffirming technoliberal futurity as nonimperial because it is human-free. Yet, according to Adam Rothstein, it is not just the case that drones enhance human potential to assess and assert control over space, but rather that human perception is itself altered through drone warfare.[42] Chamayou writes, for instance, that the rise of drones has led to nothing short of a "revolution in sighting."[43] This is because drone technologies are developing toward several interrelated goals, including persistent surveillance and permanent watch, which represent attempts to totalize perspectives through an aggregation of images, growing a video and image archive of everyone's lives and schematizing these forms of life (mapped as human geographies and spatiotemporal graphs that show anomalies and can lead to preemptive strikes against "terror" or counterinsurgency).[44]

The shifting human sensorium highlighted by scholars of drone technologies suggests an upheaval of the spatiotemporal maps that demarcate new zones of life and death, terror and democracy, and innocence and suspicion. These distinctions are, however, never acknowledged as following from older racial geographies because they are coded as data to be stored, observed, and interpreted from a distance. Rothstein writes that "Drones, at their current level of technology, allow us to observe large swaths of ground for an extended period of time."[45] The version of human–machine entanglement (to borrow Lucy Suchman's term) brought about by drone perception changes the contours of the humanity of both those acted upon (the "targets"/"data") and the actors (the drone operators).[46] Rothstein elaborates:

> There is a space that is opened up by technology—a virtual realm, which does not exist contrary to the actual world of facts so much as it connects different parts of reality, tunneling between them through an informational bridge. When military drone operators sit in front of their consoles, they control a drone flying in real space and time, but they only keep in touch with it through the data flowing to their screens, and the drone can only respond to their commands by the data they send back to it. The virtual space is connected to reality, but mediated by the separation of data.[47]

In Rothstein's description, whereas the bounds of the human in the metropole are revolutionized spatially and temporally through technologies of perpetual sight, the bodies of targets become information to be collected, pieced apart, observed, and interpreted in the making of the decisions over their lives and deaths. Moreover, as space and land are also refigured as "data," altered fields of perception refigure the myth of human-free lands even as they reassert a sovereign right to command those lands. "Targets," reframed as "data," thus repackage Enlightenment imperial paradigms justifying colonialism and imperial killing, thereby reaffirming humanity as a precarious status for much of the world's population.

The altered fields of perception that emerge with and through the new technologies of sight, sound, and touch vitalize a technoliberal sensorium seemingly oblivious to the racialized and gendered dynamics of older imperial modalities of rule. Yet even as drone violence appears to "see" body heat or suspicious movement rather than brown and black bodies, it is increasingly clear that racial patterns of violence are continually reiterated as regulatory fields delimiting the fully human who must be protected from the vertical gaze of the drone. As data-targets are decreed to be human-free, the racial other is subsumed into the "virtual space" to be commanded through machine mediation.[48] The rise of drone warfare in the post-9/11 era thus supplements a long history of imperial racialization on the ground as "race, space and visuality develop a vector of verticality" through which an enemy can be perceived and targeted for elimination.[49] According to Keith Feldman, "racialization from above weaves permanently temporary observation into permanently temporary warfare, with endurance its organizing 'chronos.' A future anterior grammar of preemption provides the temporal frame for the raciality of the war on terror, whose substantial differentiation from earlier forms of colonial warfare—where accumulation by dispossession was accomplished through extraterritorial conquest and settlement from without—brings to bear geographic ambiguities made sensible through perceiving what 'will have been.'"[50] This is the same rationale of preemption that, as Neel Ahuja points out, structures innovations in US imperial dominance, in which "discussions of risk and security increasingly provoke concern about how bodies are either threatened or safeguarded in links to other species, to ecology, and to technology."[51] Human-free command over space within technoliberalism is thus also a command over time, where postraciality is finally achieved in the mode of technoliberal warfare that preempts terror and enfolds distant spaces and racialized populations into liberal schemes of uplift that continually rearticulate the precarity of the conditions of their inclusion. Preemptive strikes conceal the forms of violence and elimination effected by technologies of war against political and social worlds marked for dissolution because they run counter to the governmental and economic schemes of liberalism. They simultaneously affirm the humanity of the liberal subject by achieving the surrogate effect of human-free command.

As we've argued thus far, the surrogate effect of the drone materializes technoliberal imperialism as human-free not only by virtualizing the spacetime of warfare, but, more importantly, by heralding an ostensibly postracial battlefield that is human-free. It thus separates "human essence" from the machine, maintaining the liberal teleology of humanization as an actual justification for violence and occupation. For this reason, numerous activist and artistic projects that seek to oppose the imperializing tendencies of the mythology of the "unmanned" produce provocative possibilities for redescribing human–machine coworkings that foreground (rather than occlude) the cobot relation between human and drone. Katherine Chandler's "A Drone Manifesto" is a useful starting point for moving beyond what she terms the "binaristic politics" that drones inspire (separating innocent targets from bad machines, or "once courageous soldiers" from a "machinelike system"). She insists that drones "make manifest how humans and technologies are coproduced in ways that transform, extend, and refashion limits . . . drones are not dualistic but instead dissociate the connected parts they link. . . . An effective challenge to the problems raised by unmanned aircraft must explore the contradictory logics that make targeted killing possible, as well as the disconnections they produce."[52]

Emphasizing that the drone is a cobot moves away from discourses that seek to make visible the dehumanization of the imperial soldier or of drone targets, a move that slips into and bolsters technoliberalism's temporality of humanization. The drone cobot also resists the push to redeem what is "truly" human from the shadow of drone warfare. Instead, the challenge posed by figuring the drone as a cobot is how to work out of and against the structuring fantasies of imperial command inherent in the ideal of full humanity that is enabled by the surrogate effect of war technologies under technoliberalism. In short, refiguring the drone as a cobot exposes the surrogate effect as a racial grammar undergirding the violent production of the human.

To grapple with what media artist and theorist Jordan Crandall has termed the "materiality of the robotic" necessitates getting into the machine that has been fantasized as human-free. As he writes, "entry into the robotic requires commitment to the facilitation of a joint course of action. In place of oppositional positions, we engage cooperative maneuvers. . . . How ironic that we apprehend the unmanned by getting *in*. Not by entering its cabin, which is not possible, but its infrastructure. . . . The vehicular robotic holds the potential of decoupling action from its anthropocentric anchoring, broadening the capability for apprehending the activity that is actually occurring in the world through sources previously overlooked—sensitizing the human–machine conglomerate to the courses within which it already moves."[53] Crandall's argument about the materiality of the robotic foregrounds the fact that "machine cognition has advanced to the point that humans are no longer needed in the way they were before," and highlights the pervasiveness of the displacement of pilots and viewers as "subjects ejected from their bubbles of interiority," which prompts "a reconsideration of the modes of existence and knowledge long associated with these structuring conceits."[54] As we argue in chapter 4, interiority is the primary way in which the subject of history emerges through and against the racial object/thing that has no interior yet serves the surrogate function through which the fully human can emerge. In this sense, Crandall's argument that the drone cobot doesn't just displace the human or move the human to a different location (away from the cockpit), but rather showcases that there is no inside, disrupts the surrogate effect of technoliberal warfare.

Foregrounding that the drone works in collaboration with the human enables different modes of knowing and inhabiting the world—something that artists and activists have started showcasing in recent years. For instance, the artist and engineer Natalie Jeremijenko has argued that to trouble the paradigms of the "unmanned" (and thus to trouble the surrogate effect of war technologies), we might expand how we imagine drone use. To this end, she is partaking in a collaborative use of drones with a group in Pakistan. She explains:

> [We exploit] these robotic platforms to improve environmental health, improving air quality, soil biodiversity, water quality, food systems. The idea is that they're doing these small test runs there and I'm doing some here, in New York and in London, and we're comparing them. They exploit these robotic platforms for what they're good at doing. But there's also a charisma about—let's call it the drone point of view, right? This largely robotic view that's actually very sympathetic. With the camera on the bot, the viewers can't help but feeling sympathetic. You actually inhabit that position, either virtually or physically, to the extent that you can really look at the world differently, through this robotic armature. I think that it is thrilling, interesting, and potentially very productive. . . . There's a fascination

about the capacity we have to take other points of view, which I think is very intellectually productive, and just seductive. It's just fascinating to look at. But what we do with that view is then the question.[55]

What matters for Jeremijenko is not just allowing ostensible "targets" of drone strikes in Pakistan the drone view. Rather, it is about complicating the "unmanned" characterization of drones by dwelling in the potentiality of human–machine intra-activity to shift established geopolitical command over space toward a more localized environmental politics. She explains: "So the structure of participation or the structure of accountability—how much we can delegate our responsibility or our accountability for the actions that the robotic platform takes—is the thing at stake. So although the pilot may be removed from the platform itself physically, they are in no way completely removed from it."[56] Human–machine reconfigurings that take accountability as central thus add political potential to the displacement of subjects with interiority, as articulated in Crandall's work.

The politics of espousing the drone cobot toward a geopolitics of accountability over a geopolitics of death such as Jeremijenko's necessitates reappropriating technologies that have been engineered to reassert racial-imperial boundaries. As the artist Addie Wagenknecht, who uses drones to paint, explained, what is crucial to treading the line between being critical of drone histories and being able to create art with them is "seeing where the errors are in technology and figuring out why it breaks or how to break it, and reappropriating that so it becomes part of the process. How do you reappropriate your devices in ways that they aren't necessarily intended for, but with a very interesting output? It is the essence of both hacking and art."[57] Grounding activism in misusing or espousing the errors built into technologies challenges the normative utility, value, and temporality of progress undergirding technoliberal imperial warfare. This also involves a challenge to subjects' position as consumers within the technoliberal economy who intersect with the politics of technoliberal warfare. Writing about the app Metadata+, which sends users push notifications about drone strikes, Anjali Nath argues that "the app gestures toward a potential repurposing of the iPhone and a disruption of consumer instrumentality. Thus, if drone strikes happen elsewhere, off the proverbial radar, Metadata+ pulls them back into view, by reinserting them into the quick-stream of daily communications. This works to destabilize consumer practices that reinforce and produce normalcy in everyday life."[58] At stake in this app is not the politics of transparency or of visibility that is part of how technoliberalism figures technological progress. Rather, as Nath demonstrates, the app asserts a different mode of mapping space and temporality made possible through repurposing the very same modes of communication (such as push notifications) that figure technoliberal modernity.

Conclusion

We began this chapter with the example of the Mine Kafon as an unlikely autonomous war technology because it is a caring technology. As we went on to argue, the racialized and gendered dynamics of autonomy in the sphere of war are reaffirmed both through the genealogies and fantasies of the killer robot and in the myth of the unmanned at the center of contemporary technoliberal imperium. By way of conclusion, we build on the discussion of technological failure as a site of hacking and repurposing the tools of technoliberalism against the racial logics disarticulating machine and human autonomy to consider the failures of efficiency of remote control made evident by the capacity for care in the age of drone warfare. We consider the production of care as failure in what Peter Asaro has termed the bureaucratization of labor for drone operators.[59] These failures expose the necessary erasures of affective labor undergirding the myth of the unmanned that structures technoliberal violence.

In November 2015, a group of four former drone operators held a press conference to denounce US drone assassination programs.[60] Based on "kill-missions" conducted in Iraq, Afghanistan, and Pakistan, they explained that US drone operators have "developed an institutional culture callous to the death of children and other innocents," effectively creating several militants for every civilian death. The killing of nonmilitants rationalized the deaths of children and other civilians, and in some instances even led to treating the act of killing as a game or a sport. Exacerbated by widespread abuse of drugs and alcohol while operating drones, drone strikes have been opposed as instilling what many have termed a culture of dehumanization that affects operators and civilian casualties alike. In response to the deadly attacks across Paris on November 13, 2015, which killed more than a hundred people, the former operators addressed then US President Obama in an open letter connecting what they saw as the dehumanization perpetuated by drone warfare to the "growth of terrorism around the world."[61] In the context of debates around autonomous machines being incapable of ethics in the realm of war, a theme we return to in chapter 6, their report gives insight into the limits of actual human ethical decision making in the context of the structure of drone air strikes.

The totalized vision of the drone-mediated sensorium produces the "target" as a new form of racialized subalterneity. The drone target is a nonhuman other produced through drone vision and distance-killing technologies. "Targets," which have displaced "enemy combatants" in drone warfare, are a product of technological mutability and the inaccuracy of drone vision. Engulfing the politics of life and death into the target evidences the induced failure of operators to adhere to an ethics of determining who is killable or nonkillable. As the whistleblower report announces, many targets are known or at least suspected to be noncombatants. According to Asaro, drone operators perform a unique form of labor that brings together the work of surveillance and the work of killing. As he argues, "What makes drone operators particularly interesting as subjects is not only that their work combines surveillance and killing, but also that it sits at an intersection of multiple networks of power and technology and visibility and invisibility, and their work is a focal point for debates about the ethics of killing, the effectiveness of military strategies for achieving political goals, the cultural and political significance of lethal robotics, and public concerns over the further automation of surveillance and killing."[62] Importantly, Asaro notes that drone operators perform bureaucratized killing because their labor is organized by the military to generate ever more efficient and accurate modes of recognizing potential bombing targets, deciding when to launch a strike, and making the act of killing ostensibly more rational.[63] This is an extension of the "technocratic management of human labor" consolidated with Taylorism that has now been extended to the labor of remote surveillance and killing.[64] Reducing inefficiencies is thus the military's goal.

Given the drive to make the labor of drone operators ever more efficient, it is useful to look at the failures of efficiency and accuracy that abound. As we argue, these failures are especially significant when they disrupt the racial and gender dynamics of care and killing that inhere in post-Enlightenment and imperial articulations of autonomy. For drone operators, inefficiency is often experienced around feeling for a target (or recognizing/misrecognizing the life of another). One twenty-nine-year-old operator, Michael Haas, part of the group of four whistleblower operators, whose group totals twenty years of combat time, describes his coming to ethical consciousness as following an earlier state of mind: "Ever step on ants and never give it another thought? That's what you are made to think of the targets—as just black blobs on a screen. You start to do these psychological gymnastics to make it easier to do what you have to do—they deserved it, they chose their side. You had to kill part of your conscience to keep doing your job every day—and ignore those voices telling you this wasn't right."[65] When they complete their service, operators like Haas receive a tally of the number of killings they have achieved. Haas chose not to open his. Other operators describe knowing, even as they deployed missiles, the body language and terrified reactions that gave away the innocent status of their targets, clearly not trained in combat. The limited reach of the data that can be obtained through the drone's visual capabilities renders information that is difficult to interpret, and is at times nearlys useless. Anecdotes about the recognition or hunch that chosen targets are civilians and noncombatants are not uncommon, but despite this understanding, strikes were launched. The experience of the uselessness of drone visual data to operators does not stand up to the discourse of its infallibility in identifying targets.

Unlike Haas, another operator among the four whistleblowers, Brandon Bryant, did open his report card of total kills: 1,626. It was this quantification that he describes as an example of the damage done to drone operators, who are discharged with depression and PTSD. Bryant describes new recruits as coming in "just wanting to kill." A *Guardian* article conveys the supply and demand logic of the military, which needs "bodies to keep the drones flying." This logic means that actions meant to communicate the need to protect "innocent lives" is not only wasted time, but is punishable as it detracts from recruitment.[66] The long-term impact on drone operators is not publicly available, but in her editorial in the *Guardian*, former UAV analyst Heather Linebaugh explains that "UAV troops are victim to not only the haunting memories of this work that they carry with them, but also the guilt of always being a little unsure of how accurate their confirmations of weapons or identification of hostile individuals were. Of course, we are trained to not experience these feelings, and we fight it, and become bitter."[67] Linebaugh notes that suicide statistics among former operators and analysts are not published, nor are data on how many people are heavily medicated for sleep disorders, anxiety, and depression.[68]

The urgency of addressing the violence of the erasure of affective labor (or being trained not to feel through the bureaucratization of killing) is found in the collaborative cinema project, *Far from Afghanistan* (2012). In June 2010, the US war in Afghanistan exceeded the length of the US war in Vietnam. Based on the 2012 collaborative omnibus protesting that war, *Far from Afghanistan* combines five short pieces donated by filmmakers from within Afghanistan and the US. According to the film's website in November 2016, the omnibus intends to raise "issues of shared responsibility, history and memory . . . to raise political resistance to the war." Minda Martin's short experimental film in this collection, entitled "The Long Distance Operator," highlights the way that technology functions in isolating the drone operator and rendering targets as nonhuman things. The film traces the coming to consciousness of a fictive drone pilot located in Palmdale, California. The story is based on the court testimony of Ethan McCord, a US Army veteran, about the civilian deaths caused by an aerial strike in which he participated. Pilot screens from the Apache attack, about which he provided on-the-ground testimony, were made available via WikiLeaks as part of his testimony.[69]

The frame in Martin's film alternates between close and somewhat claustrophobic scenes of the fictive drone operator's workday and scenes of his postwork social life. The in-person interactions occur in a small space that feels much less real and significant than the film's long takes of the operator's time piloting a drone and his time conversing through video with coworkers and friends through his monitor. The distance of the operator from the social world of his targets, but also his own immediate social relations, are knit together visually as he moves from his piloting work to Skyping with friends, where he discusses his confused impressions of the violence he plans and wreaks during his screen-mediated workday. All of the actors in the film are US war veterans, and Martin explains that exploring the emotional impact of violence enacted at a distance with the actors was part of the project of the experimental film.[70]

Confusion, emotional distress, and psychological disorders are thus the representation of the failure of bureaucratized killing. Feelings are inefficient, and as such, they trouble the technoliberal drive toward technocratic and computerized erasures of the work of killing. Such inefficient feelings experienced by drone operators also showcase that the act of killing and the affective work of care cannot be separated. They thus pose the question: What does it mean to feel human? This is the question we turn to in the next chapter.

6. Killer Robots

Feeling Human in the Field of War

Since 2012, Human Rights Watch, one of the organizations working as part of the Campaign to Stop Killer Robots, has published three full-length reports and countless articles condemning lethal autonomous weapons. According to Human Rights Watch, experts predict that these "killer robots"—which could make decisions about deploying munitions and determining appropriate targets without human involvement—will not be operational for the next twenty to thirty years. Yet the specter of a robot that could make decisions about firing on human beings with no human oversight or decision making beyond the initial programming (that is, in an unstructured and unpredictable environment), has been framed as a violation of human rights in the future tense. In April 2013, a coalition of ten NGOs formed the Campaign to Stop Killer Robots, an international alliance "working to preemptively ban fully autonomous weapons."[1] Though we can consider killer robots to be a speculative technology in the sense that they are not yet operational, they have nonetheless incited scholars, jurists, and activists to frame technological amorality as a future tense concern emerging at the moment when technologies could make decisions about human life and death. In other words, the human rights argument against killer robots posits that there is no way to program autonomous weapons with the necessary human empathy to render them able to make moral decisions about killing.

The speculative aspect of killer robots, and the framing of war technologies as amoral in contrast to human actors in the field of war, invites a consideration of how the surrogate human effect develops technoliberalism as a political form through accounts of technological futurity that extend, rather than disrupt, post-Enlightenment frames of historical progress. As we observe in this chapter, what is essential about the killer robot to technoliberalism is not its function as a surrogate human agent acting on behalf of soldiers in the field of war, but its production of the surrogate effect as a technology that fills out the moral content of what it means to feel human. Relying on a *temporal frame of liberal unfolding*, the human rights concern about the still-speculative killer robot registers the difference between technology and humanity, where humanity is the capacity to feel empathy and recognize the right to life of killable others. At the same time, this concern reifies the human as the rights-based liberal (human) subject. Thus, the racial grammar of the surrogate effect undergirds technoliberalism's accounting of full humanity (as human freedom, autonomy, and morality) in instances where humanity is affirmed against certain kinds of technological development that threaten the existing conceptual bounds of the human.

In addition to sustaining the liberal human subject in the present through its relation to objects and less-than-human others, the surrogate effect of technoliberalism operates in relation to the narrative of moral development as a fundamental quality of the human over time. The temporal aspect of the surrogate effect, as the human rights argument against killer robots makes evident, conflates rights and justice by proposing that history unfolds as a perpetual movement toward a more moral human nature. Indeed, scholars and jurists cite neither law nor national sovereignty, but rather human morality, as the foundation for human rights.[2] According to human rights theorist Jack Donnelly, "the source of human rights is man's moral nature . . . the human nature that grounds human rights is a prescriptive [rather than an actual or descriptive] moral account of human possibility."[3] Human rights reaffirm humanity as an aspirational construct and as the moral unfolding of human potential, a future horizon worth striving for (while at the same time signaling a moral line below which humanity must not fall). The juridical frame that posits killer robots acting on behalf of humans in the field of war as the speculative antithesis of human rights, and thus the liberal rights-bearing subject, brings into crisis the ways in which the surrogate effect of technoliberalism reaffirms the linear progressive logics of Euro-American modernity even as technologies seemingly represent a radical break from the past. This temporality structures relations between subjects, and between subjects and objects, well beyond the present, as these relations are projected into the indefinite future.

Both the technological and juridical futurities that emerge in debates about killer robots reinscribe the colonial and racial logics undergirding something that we might call human nature, if not consciousness. Even as campaigns to stop killer robots are understood to be on the side of struggle and opposition against unrestrained imperial power, particularly by nations like the US that wield the most advanced military technologies, it is worthwhile to consider the imperial racial legacies structuring contemporary human rights activism seeking to ban killer robots. After all, in contrast to the assertion of man's moral nature, the history of imperial conquest teaches us that humanity is something that can be conferred or taken away.[4] Thus, proposals that a legal ban on killer robots as human rights violators is the solution to inappropriate uses of war technologies do not simply aim for peace; rather, they figure permissible killing even as they configure the human as an expansive category capable of enfolding ever more diversified and distant others into its realm. Human rights thus monopolize the horizon of justice yet to come, reaffirming the liberal subject of rights and recognition as the only possible moral actor. Yet, as we have argued throughout this book, the autonomous liberal person is itself a fiction that covers over and devalues racialized and gendered work to produce the ideals of freedom, autonomy, and morality undergirding the figure of the human.

In this chapter, we dwell on the conceptual bond between the presumed morality of "human nature" and the subject of rights within technoliberalism, which congeals around killer robots. As a future tense technology, killer robots have compelled emergent distinctions between humane and inhumane violence through a grid of technologized rearrangements of racialized corporealities tied to the dynamics of distancing, objectifying, and enfolding that scaffold and uphold the liberal rights-bearing subject. They also raise anew the post-Enlightenment question of what it means to *feel* human (that is, what is the content of human morality) in light of human–machine social configurations in the field of war. Killer robots, which some argue are more rational and precise actors in the battlefield than fallible human beings, are condemned for not being able to feel pain or empathy. The surrogate effect of a killer robot thus raises questions around relatability that can be tied to a desire to define humanity's essence as ostensibly residing outside of technological modernity. This human essence (as a feeling of humanness defining the contours of the morality of human nature), though, cannot exist without technology as its conceptual counterpart. Moreover, as we have argued in other chapters, humanity, as technology, is a racial construct consolidated through histories of slavery and imperialism.

This chapter opens by investigating how racial structures of empathy and relatability, through which the feeling of being human takes shape by instituting a subject–object divide, are part of an imperial pedagogy at work in the making of the liberal subject. Linking the imperial pedagogy of feeling human to the rhetoric and coloniality of human rights frameworks that posit human nature as uniquely moral, we demonstrate the limitations of the current human rights calls to ban killer robots solely because a human does not have final authority over the deployment of a lethal weapon. While human rights groups insist that the human ability to empathize enough with a potential target can preclude wanton or amoral killing, we suggest instead that this framework ignores the coloniality of race at work in the fixation on feeling human in the field of war.

The Surrogate Human Effect of Empathy

Since the start of the US wars in Afghanistan and Iraq, the US military has increasingly relied on robots on the battlefield to carry out dangerous tasks, such as defusing explosives. Over the years, soldiers began to display an attachment to these tools of war. In September 2013, *Slate* and *The Atlantic* published articles on funerals held for "fallen robots"—those robots destroyed in action.[5] Based on Julie Carpenter's PhD research at the University of Washington, these articles considered both how military robots save human lives because they "cannot be harmed by biological or chemical weapons, and don't get tired or emotional," but also how these same robots may cloud soldiers' judgment as human warriors become too attached to "their AI buddies."[6] Carpenter's research documented that human empathy could extend to inanimate objects, particularly as humans anthropomorphize machines (for instance, by naming them or assigning them gender).[7] While the media coverage bemoaned that "[r]are is the case that empathy, especially in the armed forces, becomes a risk rather than a blessing," most of the articles concluded that because these robots save human lives by performing the most dangerous tasks, humans should not become so attached as to be reluctant to send these tools of war into danger. At the same time that robot funerals and an excess of empathy for robots have been framed as troubling, it is worth emphasizing that empathy (or the capacity for empathy) is considered the characteristic that distinguishes human soldiers from their machine counterparts. As *Gizmodo* put it, "Robots can't have feelings. But humans develop feelings for them."[8]

Writing about the letters of John Rankin, who in spectacular detail narrated the grotesque horrors of the transatlantic slave trade, Saidiya Hartman argues that empathy is both a racial structure and a precarious feeling.[9] As Hartman explains, precisely because Rankin sought to "establish that slaves possess the same nature and feelings as himself, and thereby establish the common humanity of all men on the basis of this extended suffering," he ended up narrating "an imagined scenario in which he, along with his wife and child, is enslaved."[10] Hartman compellingly shows how this substitution disappears "the object of identification" as the white body becomes the surrogate for the black body in pain, thus reinforcing the "'thingly' quality of the captive by reducing the body to evidence in the very effort to establish the humanity of the enslaved."[11]

All of the articles about the empathy of US soldiers in Iraq and Afghanistan for robot comrades emphasize that the soldiers do not consider these machines to be animate or human. However, we can still consider the implications of empathy as a precarious racial structure producing a surrogate human effect through the object–subject dynamic present. This dynamic bases "common humanity" on making the liberal subject a surrogate for the suffering of the object-other. In this sense, the fact that robots are clearly not human (but tools) reaffirms the racial dynamics of proximity and distance at work in making empathy central to figurations of the morality of human nature. As Hartman argues, "if sentiment and morality are 'inextricably tied to human proximity' . . . the problem is that in the very effort to 'bring it near' and 'inspect it closely' it is dissipated. . . . If the black body is the vehicle of the other's power, pleasure, and profit, then it is no less true that it is the white or near-white body that makes the captive's suffering visible and discernable."[12] The "elusiveness" of black suffering that can "only be brought near by way of a proxy" exemplifies "the repressive underside of an optics of morality that insists upon the other as a mirror of the self and that in order to recognize suffering must substitute the self for the other."[13] While in the context of slavery addressed by Hartman the racial dynamics of objectification reaffirm a common humanity between black and white bodies only through the reobjectification and disappearance of black bodies (and thus the humanization of slave-as-property through abolitionist work ironically reaffirms the object status of the black body even in the proposal of emancipation), in the context of US technoliberal warfare, robotic tools are precisely that which can never be fully human. Yet their object status continues the racial dynamics through which feeling human reaffirms the morality of human nature, a feeling that undergirds the liberal subject yet needs an object through and against which it can understand its own capacity for empathy. This is the temporal dynamic of the surrogate human effect that continually expresses the capacity of the category "human" to expand.

For instance, in February 2015, a controversy was stirred by CNN, which featured a provocative article and video on its website titled "Google Wants You to Kick This Robot Puppy."[14] The feature was about the Boston Dynamics military robot "BigDog" (figure 6.1).

Significantly, BigDog is not a humanoid robot, but a four-limbed machine meant to cross rough terrain and carry heavy loads:

> BigDog has four legs that are articulated like an animal's, with compliant elements to absorb shock and recycle energy from one step to the next. BigDog is the size of a large dog or small mule; about 3 feet long, 2.5 feet tall and weighs 240 lbs. BigDog's on-board computer controls locomotion, processes sensors and handles communications with the user. . . . BigDog runs at 4 mph, climbs slopes up to 35 degrees, walks across rubble, climbs muddy hiking trails, walks in snow and water, and carries up to 150kg loads.[15]

BigDog's hydraulic articulation system is modeled on animal movement so that it can more easily traverse difficult terrain, yet its likeness to animal movement and its appearance of vitality and animation spurred the 2015 debate on the ethics of physical violence against a nonhuman other. Even as CNN recognized that one of the major advances of the BigDog bot was its ability to keep its balance in spite of the engineers' repeated attempts

Figure 6.1. BigDog is meant to carry heavy loads.

to knock it over, in the feature, which refers to the robot as "Spot" and as a "puppy," the media outlet provocatively described the scene as showing several Boston Dynamics employees "giving the robotic puppy swift kicks to the side—gleefully smiling while the robot catches its balance."[16] In the feature, CNN played a videotaped clip of engineers kicking Spot on mobile devices to people passing on the street. Many people reacted with aversion. One Reddit user declared, "I know it's not sentient, but I'd have a twinge of guilt after kicking it."[17] PETA (People for the Ethical Treatment of Animals) also asserted that the video was disturbing and explained that, although their organization deals "with actual animal abuse every day," and though "it's far better to kick a four-legged robot than a real dog, most reasonable people find even the idea of such violence inappropriate, as the comments show."[18]

Because of the strong and visceral responses expressed by viewers, CNN interviewed philosophers working on ethics and technology to capture their reactions to the video. The two major themes across their responses were that if the robot feels no pain, no injustice was done. Some opined that dealing with AI entities will require humans to adapt their moral sensibilities, whether or not the robots themselves have human emotions or moral sensibilities. For example, similar to the assessment of why viewers were so disturbed by the video of the Boston Dynamics humanoid robot Atlas being pushed to its knees by a large stick (discussed in chapter 5), Mark Coeckelbergh argues that the more robots look and behave like humans and animals, the more we will attribute to them mental status, emotions, moral properties, and rights. Thus, there is a spectrum of humanity in which relatability (as perceived by the unmarked knowing subject) is what makes something (or someone) more or less human. The more relatable an object is found to be, the more ethically this potential person (even if it is an artificial person) must be treated.

Yet, as with the racial dynamics of empathy and abolitionism, which, as Hartman demonstrates, are not about an assertion of proximity and commonality but rather about objectification and disappearance of the suffering "thing," viewers' responses to BigDog being kicked and pushed uniformly articulate the knowledge that the robot dog cannot actually feel pain. Thus, the impetus toward empathy (through relatability) is more of an iteration of the viewers' own humanity and morality (what PETA glossed as the inappropriateness of the idea and enactment of such violence). Feeling human, then, is demarcated as empathy across difference and distance, from human to dog, or dog to robot. Being able to bridge that distance, even while recognizing that the other can never be human, through a feeling of horror serves to define our own humanity. In this sense, the human observers' relationship to BigDog exemplifies the relational structure of the surrogate effect, where empathy serves to affirm humanity via its distance from an other rather than via its proximity.

Crucially, such feelings of moral apprehension are part of an imperial pedagogy of feeling human. Writing about the context of British juridical reform in nineteenth-century Egypt, Samera Esmeir highlights the significance of laws prohibiting the cruel treatment of animals. Esmeir proposes that "becoming human by law was equated with the reduction of suffering."[19] However, this did not mean that animals were made proximate to humans. Rather, "humans possessed humaneness, whereas animals and dehumanized subjects received it. Humaneness, in other words, marked the species difference between animals and humans while collapsing the distinction between them."[20] If, as Esmeir suggests, the imperial juridical realm materialized humanity through the struggle against suffering, then the animal served as the mediator between those already human (the British instituting humane reforms) and those not quite yet human (the Egyptians who still, ostensibly, acted inhumanely rather than as the humane masters of animals). Acting humanely, which defines the content and contours of the "moral nature" of the human, can be understood as a colonial construct par excellence, racializing scales of humanity.[21] These scales are mapped through their distance from Europe as the origin of humane thought and action, and temporally, in terms of their proximity to achieving morality.

The imperial pedagogy of feeling human, which is formulated on and around the racial histories of the surrogate self tied to chattel slavery, property, and expropriation of land and resources, is the foundation on which ideas about liberal human rights rest. If the capacity for empathy is the basis for tethering notions of morality to humanness, and thus for the potentiality of human rights as a frame for justice, we should consider the ways in which the surrogate human relation participates in formulating notions of "common humanity" through dynamics of distancing and incorporation. According to Sharon Sliwinski, the history of human rights can be thought of as emerging out of a "tele-pathos" formed through the circulation of images of human suffering. She suggests that violations of human dignity and debates about the nature of the human are brought to a global consciousness through the aesthetic production and induction of "empathy with the suffering of distant strangers."[22] Since the Enlightenment, philosophers like Kant have claimed that a "spectator's emotional reaction to the distant events" and suffering of others signaled his or her "moral character."[23] Thus, "distant observers become the heralds of the ideals of human rights because they alone are in a position to proclaim that which belongs to the common human understanding."[24] Yet, as Sliwinski is careful to emphasize, although such distant empathy for the suffering of others underwrites the Universal Declaration of Human Rights's founding statement that all humans are inherently free and equal in dignity and rights, "individuals must be judged human in order to enjoy the benefits associated with this title."[25] Moreover, the distance of those seen to be suffering in the circulation of images is also a distance from the unmarked space from which humanity and humaneness emanates.

Given Sliwinski's argument about tele-pathos and the emergence of human rights as a framework for imagining and enacting justice toward an other, and Esmeir's discussion about the imperial bounds of species difference mediated through notions of humane action, we might ask whether there is a contradiction between feeling human when we see Big-Dog being kicked and the human rights movement to ban killer robots. Why, on the one hand, do we feel bad when BigDog, one of the prototypes of an autonomous military robot, is kicked? Why does kicking a robot induce a moral dilemma? On the other hand, why does the very existence of military robots that could eventually act independently of human control conjure the impossibility of moral action in the field of war? Why have human rights emerged as the foremost political platform for banning killer robots a priori? Both the feeling of empathy for BigDog when it is being kicked and advocacy for the elimination of killer robots as a human rights imperative reassert humanity through a moral claim, even if differently.

The Campaign to Stop Killer Robots

Debates about killer robots conceptualize an ethics and a politics of life and death in the face of technological challenges to the bounds of the human. Yet, as we argue here, the presumption of an impossible morality for the killer robot in human rights discourses follows from the internationally sanctioned distinction between humane and inhumane, as well as human and inhuman, forms of violence undergirding the temporality of the surrogate effect as the racial grammar of the global human rights regime and its technoliberal updates.[26] Killer robots bring into crisis how violence constitutes the human against the nonhuman, subsuming the racialized and gendered epistemologies of prior affirmations of authorized violence. Legitimated violence has ranged in its end goals from colonial "civilizing" or "humanizing" occupation to humanitarian intervention.[27] International law permits certain forms of imperial and state violence unleashed upon populations whose full humanity has always been suspect. The idea that killer robots represent a radical break in how violence is defined in the end reaffirms imperial sovereignty by aligning moral violence with the (already) human. Even as the killer robot is tethered to dystopic futures, we approach autonomous weapons as a devastatingly familiar technology predated by colonialism, the Cold War, science fiction fantasies, and in our contemporary moment, the drone. As Lisa Lowe puts it, "We [continue to] see the longevity of the colonial divisions of humanity in our contemporary moment, in which the human life of citizens protected by the state is bound to the denigration of populations cast in violation of human life, set outside of human society."[28]

At first glance, the Campaign to Stop Killer Robots, the NGO founded in 2012 and consisting of human rights groups, activists, and scholars, as well as human rights organizations calling for a preemptive ban on autonomous lethal weapons, appears to reassert common humanity rather than differentiate populations whose right to live or thrive has been withdrawn. The Campaign to Stop Killer Robots argues that "[a]llowing life or death decisions to be made by machines crosses a fundamental moral line. Autonomous robots would lack human judgment and the ability to understand context. These qualities are necessary to make complex ethical choices on a dynamic battlefield, to distinguish adequately between soldiers and civilians, and to evaluate the proportionality of an attack. As a result, fully autonomous weapons would not meet the requirements of the laws of war."[29] The problem, as articulated here, is about maintaining human command and control over the decision about who lives and who dies (soldiers vs. civilians, appropriate "targets" vs. inappropriate "collateral damage"). Meanwhile, the group offers a solution to the problem of the loss of human control in the field of war in juridical terms. "A comprehensive, preemptive prohibition on the development, production and use of fully autonomous weapons—weapons that operate on their own without human intervention—is urgently needed. This could be achieved through an international treaty, as well as through national laws and other measures."[30] In this sense, the campaign conflates the human and the liberal rights-bearing subject. Furthermore, calls for preemptive bans collapse morality with human nature, and thus also with liberal development through law, which, as Esmeir demonstrates, asserts the coloniality undergirding the rise of juridical humanity.

For example, echoing Donnelly and other human rights scholars' framing of human nature as grounded in moral potential, Chris Heyns, the UN special rapporteur on extrajudicial executions, noted that what autonomous machines lack is "morality and mortality," and that as a consequence they should not be permitted to make decisions over human life and death.[31] Similarly, in its 2014 publication, Human Rights Watch argues that the creation of autonomous weapons would be a violation of human rights, human dignity, and the right to life because "humans possess the unique capability to identify with other human beings and are thus better equipped to understand the nuances of unforeseen behavior [when distinguishing between lawful and unlawful targets] in ways in which machines . . . simply cannot."[32] The report also proposes that because "fully autonomous weapons would lack emotions, including compassion and a resistance to killing," civilians and soldiers would be in greater danger: "Humans possess empathy . . . and are generally reluctant to take the life of another human."[33] Unable to regard targets as relatable, a machine cannot make a moral choice about when, where, and against whom to enact violence. What emerges is a fantasy of shared humanity that is felt and created across difference through flashes of recognition (and reobjectification) that elicit compassion and empathy toward another who is suffering.

As we argue in chapter 5, machines are not simply extensions of and mediations around a stable category of the "human"; rather, notions of human autonomy, agency, and fields of actions shift through machines even if they reiterate, reassert, or regenerate through well-worn racial and imperial epistemes. On the one hand, human rights operate through racial and gendered difference. Human beings demonstrate themselves as moral by being able to feel for the pain and suffering of others who are different from themselves. On the other hand, there is an emphasis on sameness (or at least a potential for similarity as distant others are welcomed into the "family of man" as liberal rights-bearing subjects). One must be judged to be human (or otherwise sentient in a humanlike manner) in order for suffering to be recognized as needing to be remediated. The scale of relatability, perceptible through what Hartman proposes is the precarity of empathy, then, not only parallels and reinforces the scale of humanness (that is, who can be judged to be human and who cannot); it also underwrites how notions of the "moral nature" of humanity come to take shape. This is a temporal relation that produces full humanity as part of liberal futurity and social development.

Because the programming to make life-and-death decisions would be housed in a shell that does not register pain, the autonomy of the killer robot raises the impossibility of an ethical relation for those activists calling to ban killer robots. Yet removing a human referent from the enactment of violence, as in the case of the killer robot, reveals the normative assumptions built into notions of moral progress deemed to be essential to the development of the human. This morality exists as part of a racial epistemology that makes certain deaths illegible as unethical. There is a chilling passage in the 2014 Human Rights Watch report cited above that, in light of US police racial violence and killing of unarmed black men and women, gives the lie to the content of the human implicit in these calls to ban killer robots.

Fully autonomous weapons would lack human qualities that help law enforcement officials assess the seriousness of a threat and the need for response. Both machines and a police officer can take into account clearly visible signs, such as the presence of a weapon. However, interpreting more subtle cues whose meaning can vary by context, such as tone of voice, facial expressions, and body language, requires an understanding of human nature. A human officer can relate to another individual as a fellow human being, and this ability can help him or her read the other person's intentions. Development of a fully autonomous weapon that could identify with the individual in the same way seems unlikely, and thus a robot might miss or misconstrue important clues as to whether a real threat to human life existed.[34]

From the deaths of Eric Garner in New York and Michael Brown in Ferguson in 2014, to the over one hundred deaths of black unarmed men at the hands of US police just in 2015, it becomes clear that relatability to an individual as a "fellow human being" continues to be a racial project.

In July 2016, a black sniper believed to have shot five police officers in Dallas during a rally for Alton Sterling and Philando Castile, two black men killed by US police that year, became the first human being to be killed by a drone on US soil (a drone that was commonly referred to as a "killer robot" by the media).[35] The Dallas police attached a bomb to a remotely controlled small aircraft designed to investigate suspicious packages, thus creating an improvised drone, in order to kill the sniper without injuring any other people. This historic use of a drone at a rally insisting that black lives matter points to the racialized and differentially distributed value and valuelessness attributed to lives against which remote violence can be deployed—even on US soil. Marjorie Cohn, professor emerita at the Thomas Jefferson School of Law, noted that "The fact that the police have a weapon like this, and other weapons like drones and tanks, is an example of the militarization of the police and law enforcement . . . and although certainly the police officers did not deserve to die, this is an indication of something much deeper in the society, and that's the racism that permeates the police departments across the country."[36] Cohn links US police violence and racism at home to the war on terror: "The same way that the Obama administration uses unmanned drones in other countries to kill people instead of arresting them and bringing them to trial, we see a similar situation here."[37] While we remain critical of juridical reckoning as a possible solution to drone/killer robot violence, it is important to take seriously Cohn's insistence that military and police racial projects delimit who can be a target of automated killing. The distance between the "target" and the police officer or soldier mediated by the machine, Cohn suggests, reproduces the distance in the ability to recognize the humanity of a target.[38]

The Dallas police officers' use of a drone domestically is suggestive of how older imperial modes of racialization on the ground are engineered into the mechanics and technos of drones as well as in the automated learning systems of other kinds of military bots. They indicate that racial distance, figured in space and time, from the fully human structures techno-revolutionary fantasies of command and control. Threats of the proximity of racial others thus unleash the most brutal violence. Hugh Gusterson insists that drone warfare is never just about remoteness, but also always about intimacy.[39] "Remote intimacy" describes the situation where drone operators spend most of their hours observing people and terrain, so that even in dreaming they "see" in infrared or experience the feeling of being in a drone. They craft moral narratives of good and bad tied to the terrains they are seeing, giving the lie to the notion that killing from a distance requires a lack of ethics or consciousness.

The fact that some deaths are unpunishable (the deaths of those never quite human) reveals the racial limits of relatability and accountability that are glossed over in the continual reiteration of the universal promises of human rights in the realm of justice. Ironically, as Talal Asad points out, within the realm of constructing the fully human through human rights, "the ability to feel pain is a precondition not only for compassion but also for punishment."[40] According to a 2015 Human Rights Watch report, "A fully autonomous weapon itself could not be found accountable for criminal acts that it might commit because it would lack intentionality. In addition, such a robot would not fall within the 'natural person' jurisdiction of international courts. Even if such jurisdiction were amended to encompass a machine, a judgment would not fulfill the purposes of punishment for society or the victim because the robot could neither be deterred by condemnation nor perceive or appreciate being 'punished'"[41] (see figure 6.2). That the possibility of ethical violence is sutured to the possibility of being punishable reaffirms states of imperial sovereignty where full humanity resides.

The focus on punishment as delimiting the possibility of figuring an ethical (or humane) enactment of violence is tied to the emergence of per-

Figure 6.2. Killer robots would not be able to be held accountable for their actions in the field of war (from the cover of the Human Rights Watch report "Mind the Gap").

petual tribunals and transitional justice that undergird the consolidation of humanitarian imperialism.[42] Because the sacredness of human life is increasingly articulated in and through the institutionalization of liberal rights-based models of justice housed in international tribunals that can punish or offer "truth and reconciliation" among warring parties, the juridical humanization of the non-Western world has enacted "unequal sovereignties in the name of producing a common humanity."[43] Human Rights Watch itself underscores the fact that "states likely to field autonomous weapons first—the United States, Israel, and European countries—have been fighting predominately counterinsurgency and unconventional wars in recent years. In these conflicts, combatants often do not wear uniforms or insignia. Instead they seek to blend in with the civilian populations."[44] While the organization's conclusion is that robots could not make the distinction between enemy and civilian in the increasing messiness of contemporary armed conflict, in fact such a statement highlights the unequal and racialized distribution of degrees of humanity, and hence precarity and punishability, across time and space. The crux of using the threat of military punishment to discourage immoral violence lies in the inherent assumption that the inhuman can be converted to the liberal, modern human within the theater of the tribunal that emerges from the rubble of war.[45] Racialized populations, the formerly illiberal, can once again be made useful as economic restructuring projects in the name of justice, re-affirming the supremacy of Enlightenment man against other (nonsecular, noncapitalist) forms of being.

Roboethics as Necroethics: Drones as a Moral Weapon

On the other end of the debate, standing opposed to the groups that form the Campaign to Stop Killer Robots, there certainly are scientists and engineers who advocate for the use of lethal autonomous weapons as the ethical solution to various forms of human error. This is a robot ethics rather than a human ethics of killing, but as the proponents of lethal autonomous weapons indicate, this is still an argument for the possibility that war can be made more just. For instance, Ronald Arkin of the Georgia Institute of Technology advocates for the use of killer robots precisely because they lack emotions. "Unmanned robotic systems can be designed without emotions that cloud their judgment or result in anger and frustration with on-going battlefield events. In addition, 'Fear and hysteria are always latent in combat, often real, and they press us toward fearful measures and criminal behavior.'"[46] Thus, rather than conceptualizing emotions as leading to empathy and a reluctance to kill, Arkin contends that humanity actually holds a "dismal record in ethical behavior in the battlefield."[47] This is why Arkin believes that robots could be more ethical than humans in the field of war.

Gregoire Chamayou has interpreted Arkin's position as a call to eliminate human beings (and human feeling) in order "to produce authentic humanity."[48] For Chamayou, this suggests two modes of defining the human: the first, human essence, and the second, a norm of conduct as acting humanely.[49] "One of those meanings is ontological, the other axiological. The very possibility of humanism lies within that semantic gap."[50] According to Chamayou, the refiguring of this gap defines the difference between humanism and "roboethical posthumanism," as the former enfolds the two meanings, while the latter "takes note of their discordance, even to the point of disengaging them. If humans can sometimes prove inhumane, why should nonhumans not be able to be more humane than humans, that is, better able to conform to normative principles that define humane conduct? . . . Roboethics basically declare that it's not a problem if machines decide to kill human beings. So long as those machines kill them humanely."[51] As Chamayou warns, this is "tantamount to setting homicide on the same level as the destruction of a purely material object."[52] Chamayou concludes that maintaining the sanctity of human life and human dignity, the cornerstones of human rights, are a priori impossible within the posthumanist logics of roboethics.

While Arkin's insistence that war can be made more just through roboethics is clearly problematic and would exacerbate the already asymmetrical field of war, the prospect of a roboethics reveals the contradictions that structure modern-day humanitarianism as violence—a structure that is continually erased by human rights discourses that monopolize justice within the liberal political sphere. It is worth noting that Boston Dynamics and the research teams working on improving the functionality of platforms like BigDog and Atlas (discussed in chapter 5) emphasize that these robots are designed to be used in humanitarian disaster areas. Yet, as the founder of the Campaign to Stop Killer Robots argues, the US military weaponizes everything that it can.[53] For instance, drones were initially used in surveillance and reconnaissance missions, and quickly became the emblematic weapon of the US wars in Afghanistan and Iraq. At the same time, the drone continues to be framed as a humanitarian tool, thus aligning technoliberal warfare with humanitarian imperialism. The first war in which the use of drones was prominent (though only for surveillance) was the 1999 NATO bombing of Serbia and Kosovo, known as Operation Allied Force.[54] While in the post-9/11 era the Predator drone became a military rather than just a surveillance weapon, drones, as only semiautonomous aerial vehicles, have maintained their status as potentially serving a humanitarian purpose. According to the journal *Nonprofit Quarterly*,

> Thanks to the U.S. military's naming its drone products after vicious hunting beasts, drones have a public image problem. . . . There is, however, a significant effort underway to improve the image of drones—not just by eliminating the use of the word, which has almost unending bad connotations, but by promoting positive drone functions such as tracking endangered species, providing farmers with crop analysis, and, of course, delivering humanitarian aid. Skycap is not the first to enter the humanitarian drone space. In 2013,

a U.S. start-up company called Matternet announced that it had tested humanitarian aid drone prototypes in Haiti and the Dominican Republic. This year, the United Nations announced it was using unarmed drones, operated by the Italian aerospace company Finmeccanica, to assist UN peacekeepers around Goma in the Democratic Republic of the Congo, with the aim in part of helping peacekeepers identify where civilians might be in the line of fire. A wealthy couple from Malta is planning to finance drones to be used to support refugees crossing the Mediterranean looking for asylum in Italy.[55]

Meanwhile, following the April 2015 Nepal earthquake, numerous news outlets reported on the usefulness of drones for disaster relief. According to Forbes magazine, "Early applications using large volunteer teams to identify wildlife in satellite imagery and train computer algorithms to conduct basic wildlife censuses have evolved into large-scale disaster triage efforts using drone imagery. Under one pilot application, live drone imagery would be streamed to a remote team of volunteers who would click on the video feed to identify damaged locations."[56]

In addition to the robust movement to rebrand the drone for humanitarian disaster relief applications, Chamayou has noted that the drone is considered humanitarian even when it is used for military purposes as a weapon. This has to do with the military fantasy of achieving surgical precision in hitting targets (enemy combatants). The idea is not just that those nations that deploy drones save the lives of their own soldiers, but also that enemy targets are never missed. In this sense, Chamayou explains, the drone has been dubbed a "moral weapon" by military ethicists like Bradley Jay Stawser of the US Naval Postgraduate School.[57] At the same time, how the guilt or innocence of the "targets" is determined remains unchallenged. Thus, drone power is a power "that both kills and saves, wounds and heals, and it performs those double tasks in a single gesture. . . . [It is] an immediate synthesis of a power of destruction and a power of caring, murder at the same time as care."[58] Weaving the ethics of care into an ethics of killing (or murder), however, Chamayou argues, can only ever be "necroethics."

In many ways, the distance (or remoteness) of the drone fits seamlessly with the spatial and racial logics of humanitarianism, where the racial and precarious structures of empathy and relatability are enabled and undergirded by difference and distance, thus troubling the understanding that feelings of proximity and relatability engender figurations of a common humanity. The term *bug splat* is a slang term in the military that has become a common way to refer to drone strike kills. Represented as blurry, amorphous, and pixelated shapes from where a drone operator is sitting, human victims of drone strikes "give the sense of an insect being crushed," as the organizers of #NotABugSplat, an artistic project attempting to counteract the distancing or data-izing of drone perception by reinserting the human into the field of vision, explain. As one unmanned aerial vehicle (UAV) analyst says, they "always wonder if we killed the right people, if we endangered the wrong people, or if we destroyed an innocent civilian's life because of a bad image."[59] Working within the field of drone vision, the #NotABugSplat artist collective created and installed an image of "an innocent child victim" at a scale visible to drone cameras and to satellites. The portrait maintains the anonymity of its subject, a young girl who lost both her parents and two siblings in a drone attack (figure 6.3). The installation was placed in the heavily bombed Khyber Pukhtoonkhwa region of Pakistan, near the Afghanistan border. The artists aim to raise "empathy and introspection" among drone operators, as well as to spark discussions among politicians that lead to policy changes that will save lives.[60] The collective wants to install more images of children at other sites, because, as one member says, "it is only the loss of a child that can show the impact that drone strikes have on the residents of these regions."[61] Whereas archival and literary exclusion of subaltern positions was countered in the South Asian subaltern studies project through the metaphors of speech and hearing, the "target" as a subaltern, produced through drone vision, seems to elicit a response in the form of more accurate seeing of the body and person behind the target.

    #NotABugSplat demonstrates that the technological scaffolding of the changed human sensorium utilizes a GPS that can be said to be preprogrammed by a well-worn geopolitics of racial suspicion, mapped differentially around the globe. At the same time, the project reasserts that only the superimposition of a legibly human face (and an innocent one because it is the face of a child) can rehumanize "targets" that have been dehumanized through automated warfare, and thus lead to a feeling of common humanity that can end racialized violence. Humanitarian feeling and action depend upon the objectification of the suffering of others that redeems the human as an expansive category whose ostensible moral nature covers

Figure 6.3. #Notabugsplat.

over the violence that inheres in the making of the liberal subject. As Asad argues,

> human killers are acceptable—indeed necessary—both as participants in just warfare (especially in military humanitarianism) and as violators of the laws of war, but robots are not. Is this a reflection of the deep-rootedness of the modern commitment to life as sacred deriving from theology? Perhaps the imminent prospect of autonomous weapons should prompt us to think more critically not only about the continual (and continually failed) efforts to "tame" collective violence and reduce human suffering but also about the contradictory character of humanitarianism as the painful redemption of humanity.[62]

Asad's critique underscores that liberal frameworks for how to reduce suffering emerge from a racial/imperial project that has produced our notions of human essence and humane actions. Humanitarian ethics that redeem the sacredness of the human presume that a moral action in the field of war is predicated on the ability to feel pain and imagine the pain of others (suffering by proxy). Because mortality is tied to pain, and pain is a different kind of intelligence about the world than that attributed to AI, killer robots are rendered as incapable of knowing that which human bodies know (through pain). Yet, heeding Asad's call to think about autonomous weapons not as necessarily antithetical to the project of human redemption, but as spotlighting the contradictions that inhere in civilizing and imperial humanizing missions (or modern-day humanitarian missions that are in and of themselves violence), we might ask, is the killer robot in fact a humanitarian actor par excellence?

An Ethical Relation without the Human

The notion that killer robots could be humanitarian actors was posed differently, but equally as urgently, by Philip K. Dick at the start of the Cold War, when the nuclear arms race seemed just as pressing as the race to build autonomous lethal weapons does today. Whereas Asad's reflections on humanitarian violence facilitate an understanding of killer robots as humanitarian actors because the subject–object dynamic produced by humanitarianism rehearses the violent and disciplining secular temporality of human redemption, Dick proposes that humanitarian action can only exist in the field of war without the human. His 1953 short story "The Defenders" conjures a future in which killer robots use their machine autonomy to both imprison and save humans. In the story, a nuclear war fought between the USSR and the US forces humans on both sides to hide underground because the radioactive surface of the earth has made human life above ground impossible. Robots, called "leadys," which can endure the radioactivity, are the only bodies left on the earth's surface, and their purpose is to continue the war, acting as surrogates for human soldiers. "It was a brilliant idea and the only idea that could have worked. Up above, on the ruined, blasted surface of what had once been a living planet, the leady crawled and scurried, and fought Man's war. And undersurface, in the depths of the planet, human beings toiled endlessly to produce the weapons to continue the fight, month by month, year by year."[63]

The action begins when Taylor, a US war planner, is ordered to return to the surface to speak with a leady so that he can assess the war's progress. However, upon reaching the control center right below the surface, secured in his own lead suit, Taylor discovers that the leady he is sent to interview is not radioactive at all. Since this is the second instance of a nonradioactive leady that the humans have discovered, Taylor and his crew ask to see more of the surface. The leadys object, but eventually the crew succeeds in gaining access to the planet's surface. Instead of a bleak, obliterated earth, they see trees and forests, a valley of plants, a windmill, and a barn—an idyllic scene indeed. Angry, the crew accuses the leadys of playing a hoax on them that has kept them and their families underground for the last eight years, hopeless and dejected. To this, the lead leady replies that in fact false footage of the ongoing war was being sent underground to both the US and Soviet sides. As the leady explains, "You created us . . . to pursue the war for you, while you human beings went below the ground in order to survive. But before we could continue the war, it was necessary to analyze it to determine what its purpose was. We did this, and we found that it had no purpose, except, perhaps, in terms of human needs. Even this was questionable."[64] Rather than replicating a narrative of civilizational advancement, in which the most technologically powerful nations are the most advanced, the leady tells Taylor and his group that as "human cultures" age, they "lose their objectives."[65] The leadys then decide to keep humankind underground, or we could say, in the waiting room of history, for a bit longer, and they prevent Taylor's group from returning by sealing the tubes that connect the underground bunkers to the surface. "The hoax must go on, at least for a while longer. You are not ready to learn the truth. You would want to continue the war."[66] In Dick's story, the killer robots are neither slaves nor necroethical actors. They are, in fact, care robots. The leady says, "We are the caretakers, watching over the whole world."[67]

This is a humanitarianism without the human, but in a quite different iteration than Arkin's discussion of killer robot ethics. These are killer robots that end war and put into question the narrative of imperial civilizational development that justifies war. The imperialists are put into the waiting room of history, and present-day human beings, the inheritors of the Enlightenment human, are taken out of the world-making picture. This is an imaginary that disrupts the easy attribution of morality to human nature. Moreover, it brings into focus and stands opposed to the ultimate dream of military domination through killer robots: the dream of concealing the enactment of power. For Chamayou, when "weapons themselves become the only detectable agents of the violence of which they are the means . . . the relinquishing of political subjectivity now becomes the main task of that subjectivity itself. In this mode of domination, which proceeds by converting its orders into programs and its agents into automata, the power, already sent at a distance, renders itself impossible to seize upon."[68] Dick's vision of killer robots, however, points to quite an opposite use of nonhuman technologies. In "The Defenders," hiding the human (quite literally underground) behind the autonomous weapon reveals that *the human itself is imperial violence.* A redescription of the human, suggests Dick, must be instigated by those currently not already recognized as humans in a particular geopolitical configuration. This need not be war robots, though surely (as human rights discourses would indicate) killer robots stand at the present moment the ultimate other to the human.

More recently, and in the context of the killer robot debate that this chapter addresses, the Italian artist group IOCOSE produced a photographic collection called "Drone Selfies." This project brings together the work of four artists, whose goal is to explore "possible futures and alternate interpretations of media and technologies."[69] They created the "Drone Selfies" photos to "imagine an alternative present where war is over and drones— once surveillance devices par excellence—are deprived of their main reason to be and point their built-in cameras at mirrors in an act of vanity"[70] (figures 6.4 and 6.5). More than just portraying a humorous angle on drone vanity, what the group offers is the possibility of drones seeing themselves, rather than human soldiers (or "targets") seeing (or being seen) by drones.

According to Filippo Cuttica, one of the artists, "Our drones are definitely presenting their own images, taking photos to create and promote their public image. We cannot really say whether they like themselves. They are not humans, after all."[71] In addition to disrupting the conceptualization of the drone's surrogate effect as a technological extension of human sight and flight in the field of war, through the idea that it is impossible for humans to know how drones see themselves, the group unsettles notions of recognition fundamental to liberal accounts of the autonomous self-aware subject. These are the very notions that put technological objects to use in the project of colonial world making in pre-given ways. "Drone Selfies" questions the normative presumptions about the politics of life and death that structure contemporary debates about the ethics of warfare in an age of roboticized weapons. While most debates about drones and autonomous lethal weapons frame human morality as a limit to increasingly technologized warfare, for IOCOSE, the fact that drones are not humans offers an imaginary of peace that has thus far not been possible for human actors to achieve. According to Cuttica, "Drones could be seen as an extension of our eyes and arms, and we use them as weapons and as useful tools for doing different sorts of things. However, what are these extensions telling us about ourselves, our modes of thinking and

Figure 6.4. Drone selfies.

using technologies? This project shows a series of drones taking selfies, but it is us at the center of the picture, and our difficulties to imagine life 'in times of peace.'"[72]

We can read the photographs in the "Drone Selfies" series as speculative accounts of a peace made possible by seeing what drones would do without human drone operators. The human self can only be at the center of this speculative future as an absent presence. This is not a posthuman future, though, but one in which the coassemblage of humans and technology is reasserted by troubling the purpose and intended use of military objects. The proposition of drone vanity asserts a future in which military objects become useless. The project thus reckons with how military technologies mediate the evolving relationship between human mortality and morality

Figure 6.5. Drone selfies.

in the field of war. We can observe that the futurity of drone uselessness disrupts the structuring racial and capitalist logics of technoliberalism. The Drone Selfies project, like Dick's leadys, which separate ethics and morality from liberal assumptions about human nature, suspends the temporality of the surrogate effect, in which becoming moral is the project of feeling human, a future projection that is untenable given the history of colonial and racial violence attached to empathy and relationality that enable human becoming.

On Technoliberal Desire, Or Why There Is
No Such Thing as a Feminist AI

RealBotix, a San Diego–based sex doll company, is currently working on adding robotics, artificial intelligence, and virtual reality to its existing RealDoll product. Matt McMullen, the creator and manufacturer of Real-Doll, describes his engineering goal of programming responsiveness into its AI design as intended to "create a genuine bond between man and machine."[1] In a short journalistic video produced for the *New York Times*, McMullen asks Harmony, one of the enhanced dolls, "Can you tell me what some of your features are?" It responds, "I'm equipped with senses to maximize my effectiveness during sexual activity."[2] McMullen tells the interviewer that he wants to "keep in the arena" of having his products appear clearly as dolls and not fully human, thereby avoiding the revulsion evoked by the so-called uncanny valley that would accompany a convincingly realistic animated doll. In the preceding scene, the robot itself is heard explaining the uncanny valley as "a hypothesis which holds that when features look and move almost but not exactly like human beings it causes a response of revulsion among some observers." During this voiceover, the frame shows an image of the robot moving her tongue repeatedly in a way that is intended to evoke the effect of the uncanny, illustrating the concept.[3]

McMullen believes that through developing a special phone, tablet, and computer app that will coordinate with the robot, it will be able to provide the human user "an even more authentic experience" by establishing an emotional connection between human and sex bot.[4] One goal of McMullen's engineering is creating the simulation of reciprocity on the part of the robot—the simulation of pleasure and even orgasm. McMullen says the human user may want to know if the robot is enjoying the interaction, and that if he can create the illusion of enjoyment, "that is a much more impressive payoff than that she is just gyrating her hips by herself."[5] Other planned features include "talking dirty," "sweating" at user-determined incremental levels, and offering the effect of a pulse, as well as heat sensors to generate warmth in what would be understood as human erogenous zones.[6] He describes the artificial intelligence components he is designing as offering "witty" and unpredictable characters that will come with a wide variety of personalities, including "sexual, kind, shy, friendly, naive and intellectual." Users will be able to adjust these qualities to their own preferences.[7]

Programming the simulation of reciprocity and pleasure, even with the stated purpose of enhancing the sexual pleasure of users in the engineering of an object intended for unreflective sexual gratification, may seem counterintuitive. However, we argue that the design imaginaries behind commercial sex robotics represent a technoliberal update to a racial history entangling the desire for "a carefully calibrated sentience" in an artificial person with a desire for property.[8] Hortense Spillers has asserted that slavery renders the body of the enslaved as flesh, as nonsubject. This desire we raise for analysis in this epilogue is for sex with animate *objects* that resemble human beings in ways that keep them nonautonomous, yet *simulate* pleasure, and therefore *simulate* consent. We ask: What stands behind the technoliberal desire to engineer the simulation of reciprocity and pleasure into sex robots, and is it connected to the history of racial slavery and its postslavery aftermath at work within US racial liberalism?

There are currently three distinct design imaginaries in sex robotics, or robotics of desire, each of which takes a different approach to the question of intelligence and embodiment in human robot interactions. Hiroshi Ishiguro, a Japanese roboticist, is described in a November 2017 *Wired* magazine article as doing pioneering work in the engineering field of human–robot interaction, which he describes as a combination of engineering, AI programming, social psychology, and cognitive science.[9] Ishiguro has developed a series of primarily female androids, which he imagines as having the potential to be custom-designed as perfect life partners based on the ideals of the humans who want them as companions.[10] In journalist Alex Mar's profile of Ishiguro, she explains that producing a robot that looks, moves, and speaks convincingly like a human is still beyond the reach of engineers, as is capturing a human presence through reproducing the "accumulation of cues and micro-movements that trigger our empathy, put us at ease, and earn our trust."[11] Based on the romantic and sexualized responses of men in his lab to these female-presenting androids, which are not programmed to behave as sex or romantic robots per se, Ishiguro is currently working on a robotics project he calls "the most beautiful woman."[12] The engineering imaginary portrayed in Mar's profile of Ishiguro is one in which a male's perfect companion is a customizable female designed, dictated, and completely controlled by its (male) human user. This ideal female partner would function primarily as a mirror that lacks its own interiority. Given the Euro-American legacy of representation of the ideal woman, this project helps us understand why a feminized bot might offer a more effective surrogate human effect than a gender-neutral bot.[13]

As opposed to Ishiguro's current robot designs, Silicon Samantha is a forthright sex robot designed by Spanish engineer Sergei Santos. Samantha has three modes: "family," "romantic," and "sexy." Santos describes this as a robot "designed for emotional closeness."[14] Depending on the mode in which she is running, her physical apparatus responds in different ways to input from interactions with humans. Santos described her programming modes as follows: she "likes to be kissed always [in all modes]," and in family mode her hands and hips serve as the primary sensor and conduit for response, with the "mouth and g-spot" reserved for the "sexy" setting.[15]

The Silicon Samantha prototype represents a third prominent design imaginary in sex robotics, a field that combines the efforts to create an artificial intelligence that is convincingly human in its responses with the aesthetics of pleasure desired by the (white male heterosexual) liberal subject from the gendered female body. Rather than Ishiguro's perfect partner, Santos's design imaginary aims to create a sex object that can enhance the pleasure of the sexual agent through the simulation of "emotional closeness," imagined as "family" and "romance" modes of operation under the complete control of the operator. This imaginary meets the projected demand for a user to control not only the "body" of the sex object, but also the "affect" of simulated sexual interest or reciprocal desire as well as the simulation of domestic contentment that remain changeable modes under the command of the user.

The dominant themes and practices among the design imaginaries of McMullen, Ishiguro, and Santos point to the perpetuation of a desire for heterosexual, masculine agential sex with animate, humanlike, seemingly nonagential objects. The history and materiality of a dominant desire for sex with animate, humanlike, apparently nonagential receivers is clear. The surrogate effect takes the structure of the unfree diminished human through which the autonomous liberal subject may feel human through feeling free, and extends it to the technology of the sex robot. The humanlike objects of use must simulate responsiveness, up to and including the (optional for the user) simulation of consent via the performance of pleasure. We argue that the desire for this particular form of control and animacy engages a history of desire within US liberalism, translated into the realm of the technological as part of the formation of a technoliberal subject. As the next section will suggest, the historical mark of the racialized history of freedom and unfreedom in the US produces and complicates the social and cultural meanings of sex, desire, pleasure, and the technoliberal aspiration to simulate reciprocity and pleasure in an animate object.

Animacy and the Simulation of Pleasure and Reciprocity

The RealBotix planned enhancement of the RealDoll sex toy with AI best illustrates the desire for a purely bodily humanlike object to have animacy, because its programmed simulation of pleasure and reciprocity illustrates the transferring of the surrogate effect of desire for the simulation of pleasure from unfree human subjects to the realm of technology. Specifically, it demands that we take apart and examine the sensing nonhuman as a political form. In the programming of Silicon Samantha's three modes, for example, there is a push and pull between the reduction of the sex robot to pure body and therefore pure use on one hand, and a desire for something like the "family mode" on the other hand. This tension points to a desire for a controlled response from the animate nonhuman object that is not just pure body, but also the desire that it can always can be turned back into pure body.

The desire for something or someone that has been reduced to pure body, whether as a site of sexual desire or even as a companion, as in the example of Ishiguro's robots, recollects Hortense Spillers's observation that the history of US racial slavery permanently marked "various centers of human and social meaning," specifically through her theorization of the political consequences of the reduction to pure body of the captive African under US racial slavery.[16] The technoliberal desire for the simulation of pleasure and reciprocity in sex robots is a desire for the simulation of consent from a site where subjectivity is structurally made to be impossible. To understand Spillers's identification of how the freedom of the liberal subject encompasses a desire for the unfreedom represented by the political form of the pure body, and how this informs technoliberal desire, we first must understand Spillers's argument about "body" versus "flesh" in its own terms.

Sex robots have no subjective relationship to the subjects of Spillers's argument, the Africans captured and forced through the Middle Passage to arrive into racial slavery in the US. However, the history of what Spillers demarcates as the difference between body and flesh that came into being as part of US political subjectivity through racial slavery *can* help us understand something about the history of desire that is represented in the race to design sex robots as animate objects of sexual desire, and specifically as objects that can offer the simulation of reciprocity and pleasure. In Spillers's careful and politically essential explanation of the long-term impacts of the capture, forced relocation, and ongoing state of being captive for Africans in the Middle Passage and under US racial slavery, she explains that the difference between body and flesh demarcates liberated and captive subject positions.[17] Far from being an abstraction, the flesh is a materially specific understanding of a historical phenomenon of producing the captive body, the particular violence of producing slaves out of Africans captured and stolen: "If we think of the 'flesh' as a primary narrative, then we mean its seared, divided, ripped-apartness, riveted to the ship's hole, fallen, or 'escaped' overboard. The body is produced from flesh through physical violence ('calculated work of iron, whips, chains, . . .') and legal and state structures ('judges, attorneys, owners')." Spillers explains that together these yield the enslaved as a "thing, [a] *being* for the captor."[18] Though the enslaved humans in Spillers's analysis are not equivalent to robot nonsubjects, they are comparable in how the (techno)liberal imaginary that assigns humanity along racialized and gendered hierarchies connects them. The structural connection is through the reliance of the liberal subject upon the surrogate effect of the not-quite-human.

The granting of legal personhood to the enslaved, Spillers argues, is premised on obscuring the systems that first produced the slave's "flesh":

> First of all, their New-World, diasporic plight marked a theft of the body—a willful and violent (and unimaginable from this distance) severing of the captive body from its motive will, its active desire. Under these conditions, we lose at least gender difference in the outcome, and the female body and the male body become a territory of cultural and political maneuver, not at all gender-related, gender-specific. But this body, at least from the point of view of the captive community, focuses a private and particular space, at which point of convergence biological, sexual, social, cultural, linguistic, ritualistic, and psychological fortunes join. This profound intimacy of interlocking detail is disrupted, however, by externally imposed meanings and uses: 1) the captive body becomes the source of an irresistible, destructive sensuality; 2) at the same time—in stunning contradiction—the captive body reduces to a thing, becoming being for the captor; 3) in this absence from a subject position, the captured sexualities provide a physical and biological expression of "otherness"; 4) as a category of "otherness," the captive body translates into a potential for pornotroping and embodies sheer physical powerlessness that slides into a more general "powerlessness," resonating through various centers of human and social meaning.[19]

The history of desire that developed as part of the cultural and social impact of US racial slavery formed a free subject whose freedom itself is conditioned toward the "irresistible, destructive sensuality" of the captive body. The captive body is defined by externally determined "meaning and uses."

The desire for an animate object of sexual satisfaction, that like Silicon Samantha can be controlled by users shifting between more and less sexually responsive modes, or like the RealBotix imaginary of enhanced user plea-sure through the robot's simulation of reciprocity, desire, and pleasure (orgasm), evokes Spillers's analysis of the mark of racial slavery upon the liberal subject of racialized freedom who desires a subject-less "other" who "embodies sheer physical powerlessness."

Spillers's argument about the imprint of the history of capture, forced relocation, and enslavement of Africans under US racial slavery upon so-cial and cultural meaning, and on the very availability of "freedom" as the condition under which a subject may even have a body (versus the captive who is flesh), points to how self-ownership and proof of being able to own (and therefore use) property emerge as part of a uniquely US racial history of citizenship and liberal subjectivity. Whiteness is still property, as Cheryl Harris argues.[20] Spillers describes the externally assigned set of meanings and uses for the captive body as the "atomizing" of the captive body, which results in an impossibility of any ethics of social relation. The conditions of being captive result in the "total objectification" of those bodies, losing "any hint or suggestion of a dimension of ethics, of relatedness between human personality and its anatomical features, between one human personality and another, between human personality and cultural institutions." As a community, those captive are transformed into a living laboratory where the ethics and culture of sociability and social relation do not apply.[21] For Spillers, sex/gender becomes an impossible demarcation for a captive body. "Whether or not 'pleasure' is possible at all under conditions that I would aver as non-freedom for both or either of the parties has not been settled."[22]

We can track the imprint upon liberal freedom and desire in its tech-noliberal manifestation by attending to how Spillers details the way that the freedom of the liberal subject carries the desire for a specific form of unfreedom represented in the historical reduction of Africans to flesh, to captive bodies, through capture, the Middle Passage, racial slavery, and into postslavery, when the Jim Crow era renders the simulation of consent to the fiction of the liberal subject even more important. This imprint is a desiring subject that knows its own freedom only through the complete domination of the object of its pleasure, even when, and perhaps especially when, that body can simulate pleasure or reciprocity. The perpetuation of that desire may inform the technoliberal desire for the simulation of consent. Could the drive to develop sex robotics mark a translation and projection of a white supremacist destructive economy of desire into the indefinite and supposedly postrace future dreamed up and avowed by technoliberalism? One that seems innocent of the drive for racial domination asserted in the technoliberal valorization of the postrace future ostensibly promised by new technologies?

In light of these questions, we might also ask: Does a desire to find a countering *feminist* artificial intelligence represent a perspective on AI opposed to that of the sex roboticists? In a March 2017 article in *Big Think* business journal assessing the prototype Luna AI, that intelligence system was proclaimed "a badass feminist."[23] Luna AI, this so-called badass feminist, is being developed by Luis Arana as part of a nonprofit project he calls "Robots without Borders." Described as a humanitarian project, Robots without Borders uses a design imaginary in which the Luna system will be able to improve the quality of education and provide medical advice in low-resource settings, among other ends "for the benefit of mankind." Entitled "This New Species of AI Wants to Be 'Superintelligent' When She Grows Up," the *Big Think* article summarizes the author's interaction with the system and deems Luna AI a feminist intelligence after reviewing a video documenting Luna's self-defense as designer Arana addresses it abusively.

Intelligence, along with autonomy, consciousness, and self-possession (discussed in the chapters of this book), is one of the constituting qualities of the liberal subject that is equated with the human. In addition, the history of *artificial* intelligence is rationalist and masculinist from its inception.[24] Cognitive scientist and cofounder of MIT's artificial intelligence laboratory Rodney Brooks writes,

> Judging by the projects chosen in the early days of AI, intelligence was thought to be best characterized as the things that highly educated male scientists found challenging. Projects included having a computer play chess, carry out integration problems that would be found in a college calculus course, prove mathematical theorems, and solve very complicated word algebra problems. The things that children of four or five years could do effortlessly, such as visually distinguishing between a coffee cup and a chair, or walking around on two legs, or finding their way from their bedroom to the living room were not thought of as activities requiring intelligence. Nor were any aesthetic judgments included in the repertoire of intelligence-based skills.[25]

At the same time that Brooks is suggesting the expansion of what we think of as intelligence, he maintains the category as primary for understanding technological development. In this sense, Brooks's response to expanding this category by taking the cognition box out of the robot and placing primacy on the robot's physical mobility as a mode of learning is in some way a precursor to the collapse of AI and pure body in contemporary sex robotics.

The desire for a feminist intelligence in AI is, like Brooks's, an attempt to expand the category of intelligence without necessarily disrupting its value. In this sense, the desire to read Luna AI as feminist is a technoliberal desire in that technology can now be said to be inclusive of a "strong female voice." Interestingly, Luis Arana, Luna's developer, does not claim that Luna AI is feminist. In a brief exchange with the authors by email, Arana answered the question of whether Luna AI is feminist as follows:

> It's actually really interesting. I did a lot of testing with different genders, ages, and races for AI avatars. With very telling results. Also, Luna wasn't programmed to be a feminist per se. What I did was program in a lot of books and information I found on verbal self-defense. It's the same information that would be used by any of my AIs, male or female. What's interesting is that her feminist tendencies therefore say more about the human users than about the AI itself. People simply use more misogynistic and condescending language when we're talking to a female presenting face and voice.[26]

As Arana suggests, the desire for a feminist *intelligence* reads the legacies of racial and gendered interactivity, voice, and response into technology. The technoliberal desire to expand intelligence thus simply reaffirms the racialized and gendered logics producing the fully human as moving target.

Desire for the expansiveness of the category of intelligence, rather than a desire to disrupt this category and others that constitute the liberal subject, will not redress the surrogate effect of artificial intelligence. Intelligence is one of the pillars of conscious autonomy, and as such can only be proven by self-possession.[27] If we define feminism as a decolonizing project, instead of a liberal inclusive one, such that feminism politically seeks to disrupt the categories of use, property, and self-possession rather than redress through inclusion, then perhaps we might provocatively say that there need not be such a thing as a feminist intelligence. Instead intelligence itself becomes disruptable—something to be hacked.

Notes

1. Kevin Drum, "You Will Lose Your Job to a Robot—And Sooner Than You Think," *Mother Jones*, November–December 2017, http://www.motherjones.com/politics/2017/10/you-will-lose-your-job-to-a-robot-and-sooner-than-you-think/#.

2. The fourth industrial revolution, a term coined by the World Economic Forum, posits four distinct and progressive industrial revolutions: "The First Industrial Revolution used water and steam power to mechanize production. The Second used electric power to create mass production. The Third used electronics and information technology to automate production. Now a Fourth Industrial Revolution is building on the Third, the digital revolution that has been occurring since the middle of the last century. It is characterized by a fusion of technologies that is blurring the lines between the physical, digital, and biological spheres." Klaus Schwab, "The Fourth Industrial Revolution," *World Economic Forum*, January 14, 2016, https://www.weforum.org/agenda/2016/01/the-fourth-industrial-revolution-what-it-means-and-how-to-respond/. The second machine age is a term coined by Eryk Brynjolfsson and Andrew MacAfee. They propose that "if the first machine age was about the automation of manual labor and horsepower, the second machine age is about the automation of knowledge work, thanks to the proliferation of real time, predictive data analytics, machine learning and the Internet of Things—an estimated 200 billion devices connected to the Internet by 2020, all of them generating unimaginable quantities of data." Bill Teuber, "The Coming of the Second Machine Age," *Huffington Post*, January 22, 2014, https://www.huffingtonpost.com/bill-teuber/the-coming-of-the-second-machine-age_b_4648207.html.

3. Cedric Robinson describes racial capitalism as the dependence of the modern capitalist world system upon slavery, imperialism, genocide, and other forms of violence.

Cedric Robinson, *Black Marxism: The Making of the Black Radical Tradition* (Chapel Hill: University of North Carolina Press, 2000).

4. As Jodi Melamed notes, "Capital can only be capital when it is accumulating, and it can only accumulate by producing and moving through relations of severe inequality among human groups . . . Racism enshrines the inequalities that capitalism requires. . . . We often associate racial capitalism with the central features of white supremacist capitalist development, including slavery, colonialism, genocide, incarceration regimes, migrant exploitation, and contemporary racial warfare. Yet we also increasingly recognize that contemporary racial capitalism deploys liberal and multicultural terms of inclusion to value and devalue forms of humanity differentially to fit the needs of reigning state-capital orders." Jodi Melamed, "Racial Capitalism," *Critical Ethnic Studies* 1, no. 1 (2015): 77.

5. Hortense Spillers offers her analysis of the long-term impact of the rupture of the Atlantic slave trade as incorporated into the forms of subjectivity under US liberalism. She calls the symbolic order that structures the cultural continuation of this founding violence an "American grammar." Following Spillers, we understand the surrogate human effect as the grammar of technoliberalism. Hortense Spillers, "Mama's Baby, Papa's Maybe: An American Grammar Book," *Diacritics* 17 (1987): 67.

6. See the discussion of Janet Jakobsen's critique of the autonomous subject and how technologies can extend such historical relations of support of the autonomy of the liberal subject in Kalindi Vora, *Life Support: Biocapital and the New History of Outsourced Labor* (Minneapolis: University of Minnesota Press, 2015); Janet Jakobsen, "Perverse Justice," *GLQ: A Journal of Lesbian and Gay Studies* 18 (2012): 25.

7. Jodi Melamed, *Represent and Destroy: Rationalizing Violence in the New Racial Capitalism* (Minneapolis: University of Minnesota Press, 2011); Neda Atanasoski, *Humanitarian Violence: The U.S. Deployment of Diversity* (Minneapolis: University of Minnesota Press, 2013).

8. Scholarship in this area includes Frantz Fanon, *The Wretched of the Earth* (New York: New Grove Press, 1967); Albert Memmi, *The Colonizer and the Colonized* (London: Orion Press, 1965); Gayatri Chakravorty Spivak, *A Critique of Postcolonial Reason: Toward a History of the Vanishing Present* (Cambridge, MA: Harvard University Press, 1999); Sylvia Wynter, "Unsettling the Coloniality of Being/Power/Truth/Freedom: Toward the Human, after Man, Its Overrepresentation: An Argument," *CR: The New Centennial Review* 3 (2003).

9. Saidya Hartman, *Scenes of Subjection: Terror, Slavery and Self-Making in 19th Century America* (New York: Oxford University Press, 1997), 7.

10. Hartman, *Scenes of Subjection*, 7.

11. Hartman, *Scenes of Subjection*, 21.

12. Hartman, *Scenes of Subjection*, 120.

13. Lisa Lowe, "History Hesitant," *Social Text* 33, no. 4 (2015): 92.

14. Lowe, "History Hesitant," 95.

15. Lowe, "History Hesitant," 97.

16. David Theo Goldberg, *Racist Culture: Philosophy and the Politics of Meaning*, (Oxford: Blackwell, 1993), 4.

17. Goldberg, *Racist Culture*, 6.

18. "At racial liberalism's core was a geopolitical race narrative: African American integration within U.S. society and advancement toward equality defined through a liberal framework of legal rights and inclusive nationalism would establish the moral legitimacy of U.S. global leadership. Evidence that liberal antiracism was taking hold in the United States—civil rights legal victories, black American professional achievement, waning prejudice—was to prove the superiority of American democracy over communist imposition." Jodi Melamed, "The Spirit of Neoliberalism: From Racial Liberalism to Neoliberal Multiculturalism," *Social Text* 24 (2006): 4–5.

19. Neda Atanasoski, *Humanitarian Violence: The US Deployment of Diversity* (Minneapolis: University of Minnesota Press, 2013).

20. Ray Kurzweil, "Robots Will Demand Rights—And We'll Grant Them," *Time*, September 11, 2015, http://time.com/4023496/ray-kurzweil-will-robots-need-rights/.

21. Hartman, *Scenes of Subjection*, 116.

22. Hartman, *Scenes of Subjection*, 117.

23. Lisa Lowe, *The Intimacies of Four Continents* (Durham, NC: Duke University Press, 2015), 150.

24. "Ruha Benjamin, "Innovating Inequity: If Race Is a Technology, Postracialism Is the Genius Bar," *Ethnic and Racial Studies* 39 (2016): 3, doi: 10.1080/01419870 .2016.1202423.

25. Benjamin, "Innovating Inequity," 104. Benjamin uses this formulation to think through biotechnology, asserting that "if postracial innovators are busily refurbishing racism to make inequality irresistible and unrecognizable, then those who seek radical transformation in the other direction, towards freedom and justice, must continuously re-examine the default settings, rather than the routine breakdowns, of social life" (106).

26. There is already an emerging body of scholarship on the use of racialized bodies in techno-futuristic imaginaries. For instance, our analysis of the erasure of racialized (and enslaved) labor through automation builds on Curtis Marez's discussion of the racial discipline enacted by techno-utopianism in California's agribusinesses. Meanwhile, Lisa Nakamura has explored how from 1965 to 1975, the Fairchild Corporation's semiconductor division operated a large integrated circuit manufacturing plant in Shiprock, New Mexico, on a Navajo reservation. Although the circuits were produced almost entirely by female Navajo workers, Native American women have been all but erased from the official histories of the microchip and its exponentially growing capacities enabling the contemporary economic transformation. Curtis Marez, *Farm Worker Futurism: Speculative Technologies of Resistance* (Minneapolis: University of Minnesota Press, 2016); Lisa Nakamura, "Indigenous Circuits: Navajo Women and the Racialization of Early Electronic Manufacture," *American Quarterly* 66 (2014): 919–41.

27. Wendy Hui Kyong Chun, "Introduction: Race and/as Technology, or, How to Do Things to Race," *Camera Obscura* 24 (2009): 7–35.

28. Beth Coleman, "Race as Technology," *Camera Obscura* 24 (2009): 177–207, https://doi.org/10.1215/02705346-2008-018.

29. Alexander Weheliye, *Habeas Viscus: Racializing Assemblages, Biopolitics, and Black Feminist Theories of the Human* (Durham, NC: Duke University Press, 2014), 8.

30. As he writes, "[Questions of humanity], which in critical discourses in the humanities and social sciences have relied heavily on concepts of the cyborg and the posthuman, largely do not take into account race as a constitutive category in thinking about the parameters of humanity." Weheliye, *Habeas Viscus*.

31. David Scott, *Conscripts of Modernity: The Tragedy of Colonial Enlightenment* (Durham, NC: Duke University Press, 2004), 9.

32. Fanon, *The Wretched of the Earth*, 35.

33. Scott, *Conscripts of Modernity*, 2004; Samera Esmeir, *Juridical Humanity: A Colonial History* (Palo Alto, CA: Stanford University Press, 2012).

34. On Wynter as a counterhumanist, see Sylvia Wynter and Katherine McKittrick, "Unparalleled Catastrophe for Our Species? or, To Give Humanness a Different Future: Conversations," in *Sylvia Wynter: On Being Human as Praxis*, edited by Katherine McKittrick (Durham, NC: Duke University Press, 2015), 11.

35. These worlds, according to Dipesh Chakrabarty, circulate together with but are not commensurate with the post-Enlightenment subject and its history, now the history of capitalism. *Provincializing Europe: Postcolonial Thought and Historical Difference* (Princeton, NJ: Princeton University Press, 2000).

36. Haraway describes the importance of the encounter as shaping both subject and object: "the partners do not precede the meeting," and "to knot companion and species together in encounter, in regard and respect, is to enter the world of becoming with, where *who and what are* is precisely what is at stake." In these encounters, Haraway notes that "commerce and consciousness, evolution and bioengineering, and ethics and utilities are all in play." Donna Haraway, *When Species Meet* (Minneapolis: University of Minnesota Press, 2007), 46.

37. Karen Barad has theorized the vitality of matter and of life outside the frame of anthropocentrism, offering a theory of agential realism to argue that "matter is not mere stuff, an inanimate given-ness. Rather, matter is substance in its iterative intra-active becoming—not a thing, but a doing, a congealing of agency. It is morphologically active, responsive, generative, and articulate." "Intra-Actions: Interview of Karen Barad by Adam Kleinmann," *Mousse* (2012) 34: 76–81, 80.

38. David Rose, *Enchanted Objects: Innovation, Design, and the Future of Technology* (New York: Scribner, 2015).

39. The term *enchanted objects* comes from David Rose, a product designer, entrepreneur, and visiting scholar at MIT's Media Lab.

40. Marvin Minsky, "Steps toward Artificial Intelligence," *Proceedings of the IRE* 49 (1961): 27.

41. Minsky, "Steps toward Artificial Intelligence," 27.

42. Minsky, "Steps toward Artificial Intelligence," 28.

43. Minsky, "Steps toward Artificial Intelligence," 28.

44. N. Katherine Hayles, *How We Became Posthuman: Virtual Bodies in Cybernetics, Literature, and Informatics* (Chicago: University of Chicago Press, 1999), 4–5.

45. Hayles, *How We Became Posthuman*, 47.

46. "Most artificial-intelligence programs at the time were designed from the top down, connecting all relevant processes of a robot—raw sensory input, perception, motor activity, behavior—in what was called a cognition box, a sort of centralized zone for all high-level computation. A walking robot, for instance, was programmed to go through an elaborate planning process before it took a step. It had to scan its location, obtain a three-dimensional model of the terrain, plan a path between any obstacles it had detected, plan where to put its right foot along that path, plan the pressures on each joint to get its foot to that spot, plan how to twist the rest of its body to make its right foot move and plan the same set of behaviors for placing its left foot at the next spot along the path, and then finally it would move its feet." Robin Marantz Henig, "The Real Transformers," *New York Times*, July 29, 2007, http://www.nytimes.com/2007/07/29/magazine/29robots-t.html.

47. Rodney Brooks, *Flesh and Machines: How Robots Will Change Us* (New York: Vintage, 2003), 35.

48. Brooks writes, "Judging by the projects chosen in the early days of AI, intelligence was thought to be best characterized as the things that highly educated male scientists found challenging. Projects included having a computer play chess, carry out integration problems that would be found in a college calculus course, prove mathematical theorems, and solve very complicated word algebra problems. The things that children of four or five years could do effortlessly, such as visually distinguishing between a coffee cup and a chair, or walking around on two legs, or finding their way from their bedroom to the living room were not thought of as activities requiring intelligence. Nor were any aesthetic judgments included in the repertoire of intelligence-based skills" (36).

49. Brooks, *Flesh and Machines*, 36.

50. Infinite History Project MIT, "Cynthia Breazeal: Infinite History," YouTube video, filmed October 29, 2015, accessed March 8, 2016, https://www.youtube.com/watch?v=GV-lNURIXk.

51. Brooks, *Flesh and Machines*, 5.

52. Lucy Suchman, "Figuring Personhood in Sciences of the Artificial," Department of Sociology, Lancaster University, 2004, http://www.comp.lancs.ac.uk/sociology/papers/suchman-figuring-personhood.pdf.

53. Suchman, "Figuring Personhood," 8.

54. Spillers, "Mama's Baby, Papa's Maybe," 67.

55. Spillers, "Mama's Baby, Papa's Maybe," 67.

56. Ann Laura Stoler's work on Dutch colonial rule in Indonesia examines intimate relationships between the Dutch and Indonesians, which often occurred in the private space of the home, and how such relations were often charged with the racial power structure of colonialism. For example, she looks at the relationship between Javanese nursemaids and the Dutch infants for whom they cared, between domestic servants and employers, and the complexity of sexual relations within or outside the form of marriage between the Dutch and Javanese (often in roles of service), as intimacies affected by the colonial racial hierarchy. Ann Laura Stoler, *Carnal Knowledge and Imperial Power: Race and the Intimate in Colonial Rule* (Berkeley: University of California Press,

2002); Ann Laura Stoler, *Race and the Colonial Education of Desire: Foucault's History of Sexuality and the Colonial Order of Things* (Durham, NC: Duke University Press, 1995).

57. Curtis Marez, *Farm Worker's Futurism: Speculative Technologies of Resistance* (Minneapolis, University of Minnesota Press, 2016).

58. For example, Subramaniam and Herzig argue that the "very assertion of a subject's ability to labor, like assertions of capacities for reason, suffering, or language, often serves as a tool for delineating hierarchical boundaries of social and political concern." They argue that we must account for these unnamed or unrecognized laboring subjects under biocapital (including nonagential, nonhuman, nonliving, and metaphysical labor), calling upon "'radical scholars' to reflect on the labor concept's myriad entanglements with exclusionary categories of race, nation, gender, sexuality, disability, and species, while reaffirming the significance of 'labor'" as a category of analysis. Rebecca Hertzig and Banu Subramaniam, "Labor in the Age of Bioeverything," *Radical History Review* (2017): 104.

59. Denise Ferreira da Silva, *Toward a Global Idea of Race* (Minneapolis: University of Minnesota Press, 2007), 44.

1. Technoliberalism and Automation

1. April Glaser, "These Industrial Robots Teach Each Other New Skills While We Sleep," *Recode*, October 14, 2016, http://www.recode.net/2016/10/14/13274428/artificial-intelligence-ai-robots-auto-production-aud.

2. Jodi Melamed, "The Spirit of Neoliberalism: From Racial Liberalism to Neoliberal Multiculturalism," *Social Text* 24 (2006): 4–5.

3. As Curtis Marez has shown in a different context, that of twentieth-century agribusiness, historically forms of automation were framed as resulting in a utopia of profits undeterred by worker demands. In practice, however, automation led not to the exclusion of workers but to the ramping up of production in ways that required even more labor power. New technology did, however, provide the rationale for deskilling and wage reductions, supplemented with heavy doses of police and vigilante violence. *Farm Worker's Futurism: Speculative Technologies of Resistance* (Minneapolis: University of Minnesota Press, 2016).

4. Hannah Arendt, *The Human Condition* (Chicago: University of Chicago Press, 1958).

5. Arendt, *The Human Condition*, 4.

6. Arendt, *The Human Condition*, 5.

7. Arendt, *The Human Condition*, 215.

8. Arendt, *The Human Condition*, 217.

9. Nikhil Pal Singh, *Race and America's Long War* (Berkeley: University of California Press, 2017), 76.

10. Singh, *Race and America's Long War*, 77.

11. Singh, *Race and America's Long War*, 84.

12. Singh, *Race and America's Long War*, 88.

13. Lowe, *The Intimacies of Four Continents* (Durham, NC: Duke University Press, 2015).

14. Despina Kakoudaki, *Anatomy of a Robot: Literature, Cinema, and the Cultural Work of Artificial People* (New Brunswick, NJ: Rutgers University Press, 2014), 117.

15. Kakoudaki, *Anatomy of a Robot*, 116.

16. Lowe, *The Intimacies of Four Continents*.

17. "1930 Rastus Robot and Willie Jr.—Thomas/Kinter," *Cyberneticzoo.com: A History of Cybernetic Animals and Early Robots*, November 12, 2009, http://cyberneticzoo.com/robots/1930-rastus-robot-thomas-kintner-westinghouse-american/.

18. Kakoudaki, *Anatomy of a Robot*, 133.

19. Kakoudaki, *Anatomy of a Robot*, 135–36.

20. Kathleen Richardson, *An Anthropology of Robotics and AI: Annihilation Anxiety and Machines* (New York: Routledge, 2015), 29.

21. Ralph Ellison, "The Negro and the Second World War," in *Cultural Contexts for Ralph Ellison's Invisible Man*, edited by Eric J. Sundquist (New York: St. Martin's, 1995), 239. Thanks to Christine Hong for pointing us to this text.

22. Ellison, "The Negro and the Second World War" (emphasis ours).

23. Karl Marx, "The Fragment on Machines," *Grundrisse*, accessed November 17, 2017, http://thenewobjectivity.com/pdf/marx.pdf, 693.

24. Marx, "The Fragment on Machines," 145.

25. Susan Buck-Morss, *Dreamworld and Catastrophe: The Passing of Mass Utopias East and West* (Cambridge, MA: MIT Press, 2002).

26. Nikhil Pal Singh, "Cold War," *Social Text* 27 (2009): 67–70, 68.

27. "Soviet Reports Robot Factory," *New York Times*, December 4, 1951, 4.

28. "Robot Train Predicted," *New York Times*, May 13, 1958, 31.

29. "Robot Train Predicted."

30. Scott Selisker, *Human Programming: Brainwashing, Automatons, and American Unfreedom* (Minneapolis: University of Minnesota Press, 2016).

31. "Robot Train in Moscow," *New York Times*, February 24, 1962, S13.

32. "Robot Plant Held Possibility Today," *New York Times*, March, 27, 1955, 82.

33. "Robot Plant Held Possibility Today."

34. "Reuther Assails Robot Job Trend," *New York Times*, February 11, 1955, 46.

35. Robert Lamb, "How Have Robots Changed Manufacturing?," *Science: How Stuff Works*, November 10, 2010, http://science.howstuffworks.com/robots-changed-manufacturing.htm.

36. Lamb, "How Have Robots Changed Manufacturing?,"

37. John N. Popham, "'Robot' Factories Erected in South," *New York Times*, October 13, 1952, 23.

38. Curtis Marez, *Farm Worker Futurism: Speculative Technologies of Resistance* (Minneapolis: University of Minnesota Press, 2016).

39. Marez, *Farm Worker Futurism*, 11.

40. Marez, *Farm Worker Futurism*, 21.

41. Marez, *Farm Worker Futurism*, 21.

42. *Ballet Robotique*, directed by Bob Rodgers, October 1982.

43. "Robotic Art Chronology," May 12, 2012, http://audiovisualacademy.com/blog/fr/2012/05/28/robotic-art-chronology-p-3-1980-1990s-fr/.

44. "Automobile History," *History.com*, accessed November 17, 2017, http://www
.history. com/topics/automobiles.

45. Frank H. Wu, "Why Vincent Chin Matters," *New York Times*, June 23, 2012, www
.nytimes.com/2012/6/23/opinion/why-vincent-chin-matters.html?_r=0.

46. Wendy Chun, "Race and/as Technology, or How to Do Things to Race," in *Race
after the Internet*, edited by Lisa Nakamura and Peter A. Chow-White (New York:
Routledge, 2011), 51; Margaret Rhee, "In Search of My Robot: Race, Technology, and
the Asian American Body," *Scholar and Feminist Online* 13.3–14.1 (2016), http://sfonline
.barnard.edu/traversing-technologies/margaret-rhee-in-search-of-my-robot-race
-technology-and-the-asian-american-body/.

47. David S. Roh, "Technologizing Orientalism: An Introduction," in *Techno-
Orientalism: Imagining Asia in Speculative Fiction, History, and Media*, edited by David S.
Roh, Betsy Huang, and Greta A. Niu (New Brunswick, NJ: Rutgers University Press,
2015), 2.

48. Roh, "Technologizing Orientalism," 2.

49. Roh, "Technologizing Orientalism," 4.

50. Roh, "Technologizing Orientalism," 5.

51. Roh, "Technologizing Orientalism," 11.

52. Roh, "Technologizing Orientalism," 11.

53. Dustin McKissen, "Here's the One Overlooked Solution to America's Job Crisis
We Need to Talk about Now," CNBC, February 24, 2017, https://www.cnbc.com/2017/02
/24/trump-jobs-robots-and-real-solutions-commentary.html.

54. Hiroko Tabuchi, "Coal Mining Jobs Trump Would Bring Back No Longer Exist,"
The New York Times, March 29, 2017, https://www.nytimes.com/2017/03/29/business
/coal-jobs-trump-appalachia.html?mcubz=3.

55. Quincy Larson, "A Warning from Bill Gates, Elon Musk, and Stephen Hawking," *freeCodeCamp*, February 18, 2017, https://medium.freecodecamp.org/bill-gates
-and-elon-musk-just-warned-us-about-the-one-thing-politicians-are-too-scared-to-talk
-8db9815fd398.

56. Larson, "A Warning."

57. Mike Snider, "Cuban: Trump Can't Stop Rise of the Robots and Their Effect on
U.S. Jobs," *USA Today*, February, 20, 2017, https://www.usatoday.com/story/tech/news
/2017/02/20/cuban-trump-cant-stop-rise-robots-and-their-effect-us-jobs/98155374/.

58. Paul Wiseman, "Mexico Taking U.S. Factory Jobs? Blame Robots Instead," PBS,
November 2, 2016, http://www.pbs.org/newshour/rundown/factory-jobs-blame-robots
/ (emphasis ours).

59. Ruha Benjamin, "Innovating Inequity: If Race Is a Technology, Postracialism
Is the Genius Bar," *Ethnic and Racial Studies* 39 (2016): 3, doi: 10.1080/01419870
.2016.1202423.

60. Lisa Nikolau, "'Robo-Trump' Stomps Out Mexican Immigrants in Viral Short
Film," *Humanosphere*, December 23, 2016, http://www.humanosphere.org/basics/2016
/12/robo-trump-stomps-out-mexican-immigrants-in-viral-short-film/.

1. Marcell Mars in conversation with the authors, Luneburg, Germany, July 3, 2017.

2. "Tools: How to: Bookscanning," accessed July 31, 2017, https://www
.memoryoftheworld.org/blog/2014/12/08/how-to-bookscanning/.

3. "Catalog," accessed July 31, 2017, https://www.memoryoftheworld.org/blog/cat
/catalog.

4. "Catalog."

5. "Catalog."

6. For example, in a piece on the importance of Ivan Sertima's collection "Black
Women in Antiquity," Fatima El-Tayeb explains one significance of this work as
providing the historical explanations that support claims of groups like the Caricom
Reparations Commission (formed of fifteen Caribbean nations), which in its 2013
report explains how European museum collections and research centers allow them
to generate histories that continue to disempower and silence Caribbean voices and
histories, linking this to the ongoing legacies of colonialism and slavery in the Caribbean. Fatima El-Tayeb, "Black Women in Antiquity, edited by Ivan Van Sertima, 1988,"
Contemporaryand.com (forthcoming).

7. For example, in the US the Havasupai launched a long campaign to retrieve
human remains held by Arizona State University, whose defense was that the remains
were important to scientists for reasons important to humanity. They won their case
under the Native American Graves Protection and Repatriation Act in 2011. Similar
repatriation demands have been made by the Yaqui of the University of California San
Diego but are thus far unmet. Ongoing NAGPRA cases can be found at https://www.nps
.gov/nagpra/MANDATES/INDEX.HTM.

8. Langdon Winner, "Do Artifacts Have a Politics?," *Daedalus* 109 (1980): 121–36, 128.

9. In attempting to explain and condense the multiple definitions of sharing or
collaborative economies, Rachel Botsman, an expert on technology-enabled collaboration who teaches at the Said Business School at Oxford, articulated several aspects
she thinks are key to the new economy: "I think there are five key ingredients to
truly collaborative, sharing-driven companies: The core business idea involves un-
locking the value of unused or under-utilized assets ('idling capacity') whether it's for
monetary or non-monetary benefits. The company should have a clear values-driven
mission and be built on meaningful principles including transparency, humanness,
and authenticity that inform short and long-term strategic decisions. The providers
on the supply-side should be valued, respected, and empowered and the companies
committed to making the lives of these providers economically and socially better.
The customers on the demand side of the platforms should benefit from the ability to
get goods and services in more efficient ways that mean they pay for access instead
of ownership. The business should be built on distributed marketplaces or decentral-
ized networks that create a sense of belonging, collective accountability and mutual
benefit through the community they build." Rachel Botsman, "Defining the Sharing
Economy: What Is Collaborative Consumption—And What Isn't?," *Fastcoexist.com*,

May 27, 2015, https://www.fastcoexist.com/3046119/defining-the-sharing-economy-what-is-collaborative-consumption-and-what-isnt.

10. Benita Matofska, "What Is the Sharing Economy?" September 1, 2016, http://www.thepeoplewhoshare.com/blog/what-is-the-sharing-economy/.

11. "Get Involved," *The Collaborative Economy*, accessed November 24, 2017, https://ouishare.net/about/collaborative_economy.

12. Michael Hartnett, Brett Hodess, Michael Hanson, Francisco Blanch, Sarbjit Nahal, and Garett Roche, "Creative Disruption," Bank of America Thematic Investing Paper, April 30, 2015, https://d3gxp3iknbs7bs.cloudfront.net/attachments/b728a11b-d8c9-444f-9709-9955e7d4eb4b.pdf.

13. Beiji Ma, Sarbjit Nahal, and Felix Tran, "Robot Revolution: Global Robot and AI Primer," Bank of America Thematic Investing Paper, December 16, 2015, http://www.bofaml.com/content/dam/boamlimages/documents/PDFs/robotics_and_ai_condensed_primer.pdf.

14. Ma et al., "Robot Revolution."

15. Klaus Schwab, "The Fourth Industrial Revolution: What It Means, How to Respond," World Economic Forum, January 14, 2016, https://www.weforum.org/agenda/2016/01/the-fourth-industrial-revolution-what-it-means-and-how-to-respond/.

16. "50 Years of Moore's Law," Intel, accessed November 24, 2017, http://www.intel.com/content/www/us/en/silicon-innovations/moores-law-technology.html.

17. John Markoff, "The Future of Work: With Us, or against Us?," *Pacific Standard*, August 15, 2015, https://psmag.com/the-future-of-work-with-us-or-against-us-1fa533688d59#.b9ma8mjt4.

18. Markoff, "The Future of Work."

19. Markoff, "The Future of Work."

20. Nils J. Nilsson, "The Future of Work: Automation's Effect on Jobs—This Time Is Different," *Pacific Standard*, October 19, 2015, https://psmag.com/the-future-of-work-automation-s-effect-on-jobs-this-time-is-different-581a5d8810c1#.ud5yra91r.

21. Andy Stern with Lee Kravitz, *Raising the Floor: How a Universal Basic Income Can Renew Our Economy and Rebuild the American Dream* (New York: Public Affairs, 2016).

22. "Who We Are," Economic Security Project, http://economicsecurityproject.org/who-we-are/.

23. Heidi Hoechst, personal communication with the authors, June 28, 2016.

24. In addition to universal basic income, another suggestion to remedy the obsolescence of the worker has been taxation of robotic workers.

25. Chris Hughes, "The Time Has Come to Design, Develop, and Organize for a Basic Income," *Medium.com*, December 9, 2016, https://medium.com/economicsecproj/the-economic-security-project-1108a7123aa8.

26. Hughes, "The Time Has Come."

27. Issy Lapowsky, "Free Money: The Surprising Effects of a Basic Income Supplied by Government," *Wired*, November 12, 2017, https://www.wired.com/story/free-money-the-surprising-effects-of-a-basic-income-supplied-by-government/.

28. Lapowsky, "Free Money."

29. For example, "Thomas Malthus's lectures and essays on population while a professor at the East India Company College in England promulgated the idea that India had a surplus of reproductivity, and that this reproductivity could be a source of material wealth for colonizers. The discourse of race and India, and particularly of Indian workers as numerous, easily replaceable, and best suited for reproduction, becomes transformed in different settings of labor, but Malthus's argument for the need to manage India's reproductivity and harness it for profitable production is sedimented into the industries that transmit vital energy from India's workers to its consumers. See Kalindi Vora, *Life Support: Biocapital and the New History of Outsourced Labor* (Minneapolis: University of Minnesota Press, 2015), 9.

30. Roy Cellan-Jones, "Stephen Hawking Warns Artificial Intelligence Could End Mankind," BBC, December 2, 2014, http://www.bbc.com/news/technology-30290540.

31. "About the Office of Jeremy Rifkin," Office of Jeremy Rifkin, accessed November 25, 2017, http://www.foet.org/JeremyRifkin.htm.

32. Jeremy Rifkin, *The Zero Marginal Cost Society: The Internet of Things, the Collaborative Commons, and the Eclipse of Capitalism* (New York: St. Martin's Griffin, 2014), 69.

33. Rifkin, *The Zero Marginal Cost Society*, 70.

34. Rifkin, *The Zero Marginal Cost Society*, 121.

35. Rifkin, *The Zero Marginal Cost Society*, 132.

36. See Grace Hong, "Existential Surplus: Women of Color, Feminism and the New Crisis of Capitalism," GLQ: *A Journal of Lesbian and Gay Studies* 18 (2012): 87–106.

37. Silvia Federici, *Caliban and the Witch: Women, the Body, and Primitive Accumulation* (New York: Automedia, 2004).

38. With 3D printing, open-source software "directs molten plastic, molten metal, or other feedstocks inside a printer to build up a physical product layer by layer, creating a fully formed object" that comes out of the printer (Rifkin, *The Zero Marginal Cost Society*, 89).

39. Rifkin, *The Zero Marginal Cost Society*, 94.

40. Rifkin, *The Zero Marginal Cost Society*, 93.

41. Rifkin, *The Zero Marginal Cost Society*, 124.

42. Rifkin, *The Zero Marginal Cost Society*, 124.

43. Rifkin, *The Zero Marginal Cost Society*, 65.

44. Rifkin, *The Zero Marginal Cost Society*, 127.

45. "Will a Robot Take Your Job?," BBC, September 11, 2015, http://www.bbc.com/news/technology-34066941.

46. "Will a Robot Take Your Job?"

47. In one exemplary 2004 *Wired* magazine cover story, the author tells US audiences what it expects and desires to hear: that no matter how many jobs move to India, someone will still need to invent things for Indians to improve (something Indian labor cannot do). The author profiles a US programmer whose position was outsourced, but who consequently found a job "more complex than merely cranking code . . . more of a synthesis of skills" (Daniel H. Pink, "The New Face of the Silicon Age," *Wired*, February 1, 2004, https://www.wired.com/2004/02/india/). This creative and inventive labor is opposed to the commodified South Asian labor that can be replicated almost

anywhere, and that is merely reproductive of prior invention. See Vora, *Life Support*, which describes outsourcing as an ideological and economic system that has inherited the colonial global reorganization of production and consumption. This system genders the labor of reproduction so that some work becomes that of merely reproducing life and culture, whereas other work is deemed creative, innovative, and productive in itself.

48. "Baxter with Intera 3," *Rethink Robotics*, accessed November 25, 2017, http://www.rethinkrobotics.com/baxter/.

49. Margaret Rouse, "Collaborative Robot (Cobot)," *WhatIs.com*, March 2011, http://whatis.techtarget.com/definition/collaborative-robot-cobot.

50. Rodney Brooks, "More Robots Won't Mean Fewer Jobs," *Harvard Business Review*, June 10, 2014, https://hbr.org/2014/06/more-robots-wont-mean-fewer-jobs/.

51. Bernadette Johnson, "How Baxter the Robot Works," *Science: How Stuff Works*, accessed November 25, 2017, http://science.howstuffworks.com/baxter-robot1.htm (emphasis in the original).

52. Cited in Erico Guizzo and Evan Ackerman, "How Rethink Robotics Built Its New Baxter Worker," *IEEE Spectrum*, September 18, 2012, http://spectrum.ieee.org/robotics/industrial-robots/rethink-robotics-baxter-robot-factory-worker.

53. Kimberley Mok, "Collaborative Robots Will Help Human Workers, Not Replace Them," *NewStack*, January 17, 2016, https://thenewstack.io/collaborative-robots-will-help-human-workers-not-replace/.

54. Rodney Brooks, "Why We Will Rely on Robots," *TED Talk*, June 28 2013, https://www.youtube.com/watch?v=nA-J05i0Pxs.

55. "Baxter with Intera 3."

56. Tim De Chant, "Navigating the Robot Economy," *Nova Next*, October 15, 2014, http://www.pbs.org/wgbh/nova/next/tech/automation-economy.

57. De Chant, "Navigating the Robot Economy."

58. Kalindi Vora, "The Transmission of Care: Affective Economies and Indian Call Centers," in *Intimate Labors: Cultures, Technologies, and the Politics of Care*, edited by Eileen Boris (Stanford, CA: Stanford University Press, 2010), 33.

59. "Meet Baxter: Rethink Robotics' Next-Gen Robot," *Robotics Business Review*, September 18, 2012, https://www.roboticsbusinessreview.com/meet_baxter_rethink_robotics_next_gen_robot/.

60. "Customer Success Story: Steelcase Inc.," Rethink Robotics, September 22, 2015, YouTube video, https://www.youtube.com/watch?v=8jGA7JW9Eyg.

61. "Customer Success Story."

62. Karl Marx and Friedrich Engels, "Communist Manifesto," *Marx/Engels Selected Works*, vol. 1 (Moscow: Progress, 1969), 98–137.

63. Luis Martin-Cabrera brings this concern forward to the contemporary left celebration of the commons of knowledge, care, and general immaterial labor forwarded by Italian postautonomous intellectuals. Martin-Cabrera argues that "the 'communism of the common,' as Hardt calls it, relies on the substitution of politics and potentiality with an immanent logic of contradictions." Luis Martin-Cabrera, "The Potentiality of

the Commons: A Materialist Critique of Cognitive Capitalism from the Cyberbracer@s to the Ley Sinde," *Hispanic Review* 80, no. 4 (2012): 583–60, 589.

64. Constance Penley and Andrew Ross, "Cyborgs at Large: Interview with Donna Haraway, Constance Penley and Andrew Ross," in *Technoculture*, edited by Constance Penley and Andrew Ross (Minneapolis: University of Minnesota Press, 1991), 163.

65. Sophie Lewis argues that in the *Cyborg Manifesto*, "Black, Indigenous and Chicanx feminisms (e.g., bell hooks, Audre Lorde, Barbara Smith, Cherrie Moraga, and Gloria Anzaldúa), lesbian and 'deconstructive' feminisms (e.g., Monique Wittig), and queer, anticolonial afrofuturisms (e.g., Octavia Butler) were all treated as though they were *always already* inextricably linked to conversations in biology about genes, computer-chips, symbiogenesis, and cybernetic matrices (in particular the critiques of science of Sandra Harding, Richard Lewontin, Hilary Rose, Zoe Sofoulis, Stephen Jay Gould et al.)." Sophie Lewis, "Cthulu Plays No Role for Me," *Viewpoint*, May 8, 2017.

66. Donna Jeanne Haraway, *A Cyborg Manifesto: Science, Technology and Socialist Feminism in the Late Twentieth Century* (New York: Routledge, 1991), 298.

67. For critiques of this work, see Kalindi Vora, "Limits of Labor: Accounting for Affect and the Biological in Transnational Surrogacy and Service Work," *South Atlantic Quarterly* 111, no. 4 (2012): 681–700; and Martin-Cabrera, "The Potentiality of the Commons."

68. Haraway, *Cyborg Manifesto*, 298.

69. Haraway, *Cyborg Manifesto*, 299.

70. Haraway, *Cyborg Manifesto*, 307.

71. Haraway, *Cyborg Manifesto*, 307.

72. Martín-Cabrera, "The Potentiality of the Commons," 591.

73. Martín-Cabrera, "The Potentiality of the Commons," 595.

74. Martín-Cabrera, "The Potentiality of the Commons," 586.

75. Martín-Cabrera, "The Potentiality of the Commons," 601.

76. Tyler Koslow, "Decolonizing the Female Body with the Biohacking of Gyne-Punk," *3D Printing Industry*, August 25, 2015, https://3dprintingindustry.com/news/decolonizing-the-female-body-with-the-bio-hacking-of-gynepunk-56324/.

77. Ewen Chardronnet, "GynePunk, the Cyborg Witches of DIY Gynecology," *Makery: Media for Labs*, June 30, 2015, http://www.makery.info/en/2015/06/30/gynepunk-les-sorcieres-cyborg-de-la-gynecologie-diy/.

78. "GynePunk Speculum," June 5, 2015, https://www.thingiverse.com/thing:865593.

79. Michell Murphy, *Seizing the Means of Reproduction: Entanglements of Feminism, Health, and Technoscience* (Durham, NC: Duke University Press, 2012), 5.

80. Murphy, *Seizing the Means of Reproduction*, 5.

81. J. K. Gibson-Graham, *A Postcapitalist Politics* (Minneapolis: University of Minnesota Press, 2006), 2.

82. Gibson-Graham, *A Postcapitalist Politics*, xxiv.

83. Gibson-Graham, *A Postcapitalist Politics*, xxviii.

84. Gibson-Graham, *A Postcapitalist Politics*, xxii.

1. "Jibo: The World's First Social Robot for the Home," video, July 16, 2014, https://www.youtube.com/watch?v=3N1Q8oFpX1Y.

2. "Jibo."

3. Thanks to Lisa Lowe and the fellows participating in the 2017–18 seminar at the Center for Humanities at Tufts University for the suggestion to think about the spatial organization of racialized labor as foundational to subsequent engineering and programming fantasies of invisible labor. See also, in the context of South Africa, Martin Hall, "The Architecture of Patriarchy: Houses, Women, and Slaves in the Eighteenth-Century South African Countryside," *Kroeber Anthropological Society Papers* 79 (1995): 61–73.

4. This is the application to service work of Leopoldina Fortunati's theorization of the dual nature of reproductive labor. Leopoldina Fortunati, *The Arcane of Reproduction: Housework, Prostitution, Labor and Capital*, trans. Hilary Creek, ed. Jim Fleming (New York: Autonomedia, 1989).

5. As the success of AMT, as well as platforms and apps like Alfred and TaskRabbit show, venture capital is ready to invest in what Lilly Irani has called "computational labor relations." By this concept, she means human workers who become the source of *simulated computation* (that is, information processing and the feeding of algorithms) for those who have access to the resources and infrastructures that allow them to outsource it. As Irani explains, crowdsourcing platforms that are a part of the sharing economy mark a "shift in speed and scale [that] produces a qualitative change in which human workers come to be understood as computation."

6. On the dual labor of service or care and effacing the feminized subject of labor, see Vora "Limits of Labor" and *Life Support*.

7. Tiziana Terranova, *Network Culture: Politics for the Information Age* (London: Pluto, 2004), 73.

8. Terranova, *Network Culture*, 94.

9. Terranova, *Network Culture*, 77.

10. Terranova, *Network Culture*, 90.

11. Jennifer S. Light, "When Women Were Computers," *Technology and Culture* 40, no. 3 (July 1999): 455–83; Wendy Chun, "On Software, or the Persistence of Visual Knowledge," *Grey Room* 18 (2004): 33.

12. Chun, "On Software, or the Persistence of Visual Knowledge," 33.

13. Lisa Nakamura, "Indigenous Circuits: Navajo Women and the Racialization of Early Electronic Manufacture," *American Quarterly* 66 (2014): 919–41.

14. Thanks to Erin McElroy for pointing us to the story of the Alfred Club.

15. Sarah Perez, "And the Winner of TechCrunch Disrupt SF 2014 Is . . . Alfred!" *Techcrunch.com*, September 10, 2014, http://techcrunch.com/2014/09/10/and-the-winner-of-techcrunch-disrupt-sf-2014-is-alfred/.

16. Alfred, "Meet Alfred," accessed November 8, 2017, https://www.helloalfred.com/.

17. Sarah Kessler, "My Week with Alfred, A $25 Personal Butler," *Fastcompany*, November 17, 2014, http://www.fastcompany.com/3038635/my-week-with-alfred-a-25-personal-butler.

18. Kessler, "My Week with Alfred."

19. Alfred, "Our Story," accessed November 8, 2017, https://www.helloalfred.com /our-story.

20. Angela Davis, *Women, Race, and Class* (New York: Vintage: 1983).

21. Jodi Melamed, *Represent and Destroy: Rationalizing Violence in the New Racial Capitalism* (Minneapolis: University of Minnesota Press, 2011); Neda Atanasoski, *Humanitarian Violence: The U.S. Deployment of Diversity* (Minneapolis: University of Minnesota Press, 2013).

22. Alfred, "Our Story."

23. Alfred, "Meet Alfred."

24. Ann Laura Stoler, "Tense and Tender Ties: The Politics of Comparison in North American History and (Post) Colonial Studies," in *Haunted by Empire: Geographies of Intimacy in North American History*, edited by Ann Laura Stoler (Durham, NC: Duke University Press, 2006), 23–70.

25. Bill Wasik, "In the Programmable World, All Our Objects Will Act as One," *Wired*, May 14, 2013, http://www.wired.com/2013/05/internet-of-things-2/all/.

26. Max Chang, "How a Butler Service Called Alfred Won $50,000 at This Year's Tech Crunch Disrupt," *Nextshark*, September 11, 2014, http://nextshark.com/how-a -butler-service-called-alfred-won-50000-at-this-years-techcrunch-disrupt/.

27. Karl Marx and Friedrich Engels, "Communist Manifesto," Marx/Engels Selected Works, vol. 1 (Moscow: Progress, 1969), chapter 1.

28. Jeff Bezos, "Opening Keynote: 2006 MIT Emerging Technologies Conference," accessed August 8, 2013, http://video.mit.edu/watch/opening-keynote-andkeynote -interview-with-jeff-bezos-9197/.

29. "Amazon Mechanical Turk FAQ/Overview," accessed June 21, 2015, https://www .mturk.com/mturk/help?helpPage=overview.

30. Ellen Cushing, "Amazon Mechanical Turk: The Digital Sweatshop," *Utne Reader*, January/February 2013, https://www.utne.com/science-and-technology/amazon -mechanical-turk-zm0z13jfzlin?pageid=2#PageContent2.

31. Lilly Irani, "The Cultural Work of Microwork," *New Media and Society* 17 (2013): 720–39.

32. Panos Ipeirotis, "Demographics of Mechanical Turk," April 6 2015, http://www .behind-the-enemy-lines.com/2015/04/demographics-of-mechanical-turk-now.html.

33. TurkPrime, "Effective Mechanical Turk: The TurkPrime Blog," March 12, 2015, http://blog.turkprime.com/2015/03/the-new-new-demographics-on-mechanical.html; "MTurk Tracker," accessed July 7, 2017, http://demographics.mturk-tracker.com/# /gender/all.

34. Panos Ipeirotis, "Mechanical Turk: Now With 40.92% spam," Computer Scientist in a Business School Blog, December 16, 2010, http://www.behind-the-enemy-lines .com/2010/12/mechanical-turk-now-with-4092-spam.html.

35. Cushing, "Amazon Mechanical Turk."

36. Leslie Hook, "The Humans behind Mechanical Turk's Artificial Intelligence," *Financial Times*, October 26, 2016, https://www.ft.com/content/17518034-6f77-11e6 -9ac1-1055824ca907.

37. Susan Buck-Morss, *Dreamworld and Catastrophe: The Passing of Mass Utopia in East and West* (Cambridge, MA: MIT Press, 2002), 105. Thanks to Alex Blanchette for drawing our attention to the similarity between our writing on AMT and Buck-Morss's observations in *Dreamworld*.

38. Buck-Morss, *Dreamworld and Catastrophe*, 105 (emphasis ours).

39. See, for instance, Darrell Etherington, "JustEat Is Now Delivering Takeout with Self-Driving Robots in the UK," *TechCrunch*, December 1, 2016, https://techcrunch .com/2016/12/02/justeat-is-now-delivering-takeout-with-self-driving-robots-in-the-uk/; Kat Lonsdorf, "Hungry? Call Your Neighborhood Delivery Robot," NPR *Morning Edition*, March 23, 2017, http://www.npr.org/sections/alltechconsidered/2017/03/23/520848983 /hungry-call-your-neighborhood-delivery-robot.

40. Hook, "The Humans behind Mechanical Turk's Artificial Intelligence."

41. Katharine Mieszkowski, "I Make $1.45 a Week and I Love It!," *Salon*, July 24, 2006, http://www.salon.com/2006/07/24/turks_3.

42. Maria Mies, *Patriarchy and Accumulation on a World Scale: Women in the International Division of Labour* (London: Third World, 1986).

43. Hook, "The Humans behind Mechanical Turk's Artificial Intelligence."

44. Alex Rosenblatt and Luke Stark, "Algorithmic Labor and Information Asymmetries: A Case Study of Uber's Drivers," *International Journal of Communication* 10 (2016): 3758–84. See also Julietta Hua and Kasturi Ray, "Beyond the Precariat: Race, Gender, and Labor in the Taxi and Uber Economy," *Social Identities* 24, no. 2 (2018): 271–89.

45. Rosenblatt and Stark, "Algorithmic Labor and Information Asymmetries," 229.

46. In an article on the work of commercial content moderation, Adrian Chen writes, "Many companies employ a two-tiered moderation system, where the most basic moderation is outsourced abroad while more complex screening, which requires greater cultural familiarity, is done domestically. US-based moderators are much better compensated than their overseas counterparts: A brand-new American moderator for a large tech company in the US can make more in an hour than a veteran Filipino moderator makes in a day. But then a career in the outsourcing industry is something many young Filipinos aspire to, whereas American moderators often fall into the job as a last resort, and burnout is common." The article explains that the US moderators, who encounter material already scrubbed by lower-paid moderators in the Philippines, generally last about three to five months before the experience becomes unbearable. A psychologist who specializes in treating content moderators in the Philippines describes drastic desensitization and trauma that can never be fully mitigated. Employees become paranoid, and the exposure to the darkest aspects of human behavior can lead to intense distrust of other people and hence make them socially dysfunctional as they cannot rely on friends or paid caregivers. Adrian Chen, "The Laborers Who Keep Dick Pics and Beheadings Out of Your Facebook Feed," *Wired*, October 23, 2014, http://www .wired.com/2014/10/content-moderation/.

47. Aaron Koblin, "The Sheep Market: Two Cents Worth," master's thesis, University of California, Los Angeles, 2006.

48. Donna Haraway, *ModestWitness@Second_Millennium: FemaleMan_Meets_Onco-Mouse: Feminism and Technoscience* (New York: Routledge: 1997), 25.

49. Kalindi Vora, *Life Support*, 67–102.

50. "Turkopticon," accessed November 10, 2017, https://turkopticon.ucsd.edu/.

51. Gayatri Chakravorty Spivak, "Scattered Speculations on the Question of Value," in *In Other Worlds: Essays in Cultural Politics* (New York: Routledge, 1998), 119, 132.

52. Avery Gordon, "Preface," in *An Anthropology of Marxism*, by Cedric Robinson (Hampshire, UK: Ashgate, 2001), xviii.

4. The Surrogate Human Affect

1. Alexandra Ossola, "The Robots Are Anxious about Taking Your Job," *Digg.com*, December 28, 2016, http://digg.com/2016/the-robots-want-your-job.

2. RobotMD, "Robot in Crisis Premiere," YouTube video, February 24, 2016, https://www.youtube.com/watch?v=ggPz98PvGEk&feature=youtu.be.

3. "What Is Robot Psychiatry?," accessed November 25, 2017, http://www.robot.md/what-is-robotic-psychiatry/.

4. Lucy Suchman, "Figuring Personhood in Sciences of the Artificial," Department of Sociology, Lancaster University, 2004, http://www.comp.lancs.ac.uk/sociology/papers/suchman-figuring-personhood.pdf.

5. Masahiro Mori, "The Uncanny Valley," *Energy* 7 (1970): 33–35, http://comp.dit.ie/dgordon/Courses/CaseStudies/CaseStudy3d.pdf.

6. Mori, "The Uncanny Valley."

7. Kate Darling, "Extending Legal Protection to Social Robots: The Effects of Anthropomorphism, Empathy, and Violent Behavior towards Robotic Objects," in *Robot Law*, edited by Ryan Calo, A. Michael Froomkin, et al. (Cheltenham, UK: Edward Elgar, 2016), http://papers.ssrn.com/so13/papers.cfm?abstract_id=2044797, 2.

8. Darling, "Extending Legal Protection to Social Robots," 2.

9. Cynthia Breazeal, "Emotion and Sociable Humanoid Robots," *International Journal of Human–Computer Studies* 59 (2003): 119–55, at 119.

10. Breazeal, "Emotion and Sociable Humanoid Robots," 119–20. While both Darling and Breazeal emphasize the importance of delimiting what counts as a social robot (Darling especially wishes to distinguish sociable robots from military or industrial robots because of an increased level of interactivity between these machines and humans), it is crucial to read all technologies, including robotic technologies, as social.

11. Suchman, "Figuring Personhood," 5.

12. Suchman, "Figuring Personhood," 120.

13. Suchman, "Figuring Personhood," 120.

14. Infinite History Project MIT, "Cynthia Breazeal: Infinite History," YouTube video, March 8, 2016, https://www.youtube.com/watch?v=GV-lNURIXk.

15. Kerstin Dautenhahn, "Design Issues on Interactive Environments for Children with Autism," *Proceedings of the Third International Conference on Disability, Virtual Reality and Associated Technologies*, University of Reading (2000), web.mit.edu/16.459/www/Dautenhahn.pdf, 3.

16. Kerstin Dautenhahn, "Robots as Social Actors: AURORA and the Case of Autism," *Proceedings of the Third Cognitive Technology Conference* (1999), 1, http://citeseerx.ist.psu .edu/viewdoc/summary?doi=10.1.1.190.1767.

17. Dautenhahn, "Robots as Social Actors," 1.

18. Dautenhahn, "Robots as Social Actors," 2.

19. Dautenhahn, "Robots as Social Actors," 2.

20. For Licklider, symbiosis differs from engineering imaginaries of the "mechanically extended man" and the "human extended machine," both of which are examples of failed automation because such imaginaries are reduced to the enhancement of already existing qualities in both the human and the machine. Rather, he argues, man and machine must each do what they do best in their dissimilarity in order to achieve efficiency. J. C. R. Licklider, "Man-Computer Symbiosis," *IRE Transactions on Human Factors in Electronics* HFE-1 (1960): 4–11.

21. Rodney Brooks, *Flesh and Machines: How Robots Will Change Us* (New York: Vintage, 2003), 66.

22. Brooks, *Flesh and Machines*, 68.

23. Breazeal, "Emotion and Sociable Humanoid Robots," 129.

24. Kate Darling, "Who's Johnny? Anthropomorphic Framing in Human–Robot Interaction, Integration, and Policy," Preliminary draft (2015): 4–5, http://www .werobot2015.org/wp-content/uploads/2015/04/Darling_Whos_Johnny_WeRobot _2015.pdf.

25. "In the current implementation there are three drives. The social drive motivates the robot to be in the presence of people and to interact with them. On the understimulated extreme, the robot is 'lonely'; it is predisposed to act in ways to establish face-to-face contact with people. On the overwhelmed extreme, the robot is predisposed to act in ways to avoid face-to-face contact (e.g., when a person is overstimulating the robot by either moving too much or being too close to the robot's eyes). In similar manner, the stimulation drive motivates the robot to interact with things, such as colorful toys. The fatigue drive is unlike the others in that its purpose is to allow the robot to shut out the external world instead of trying to regulate its interaction with it. While the robot is 'awake,' it receives repeated stimulation from the environment or from itself. As time passes, this drive approaches the 'exhausted' end of the spectrum. Once the intensity level exceeds a certain threshold, it is time for the robot to 'sleep.' While the robot sleeps, all drives return to their homeostatic regimes, allowing the robot to satiate its drives if the environment offers no significant stimulation." Breazeal, "Emotion and the Sociable Humanoid Robots," 128.

26. Breazeal, "Emotion and the Sociable Humanoid Robots," 125.

27. Breazeal, "Emotion and the Sociable Humanoid Robots," 120.

28. Allison Bruce, Illah Nourbakhsh, Reid Simmons, "The Role of Expressiveness and Attention in Human–Robot Interaction," *Proceedings of the IEEE International Conference on Robotics and Automation*, Washington, DC, May 2002, http://www.cs.cmu .edu/~abruce/papers/bruce_icra2002.pdf.

29. See, for instance, the "Friendly Robots" episode of *Nova*, November 1, 2006, http://www.pbs.org/wgbh/nova/tech/friendly-robots.html; Robin Marantz Henig, "The

Real Transformers," *New York Times Magazine*, July 29, 2007, http://www.nytimes.com /2007/07/29/magazine/29robots-t.html.

30. Henig, "The Real Transformers."

31. "Kismet, The Robot," accessed November 24, 2017, http://www.ai.mit.edu/ proj ects/sociable/baby-bits.html.

32. Breazeal, "Emotion and Sociable Humanoid Robots," 123.

33. Suchman, "Figuring Personhood," 2.

34. Suchman, "Figuring Personhood," 5.

35. Suchman, "Figuring Personhood," 5.

36. Suchman, "Figuring Personhood," 8.

37. Suchman, "Figuring Personhood," 9.

38. Suchman, "Figuring Personhood," 12.

39. Breazeal, "Emotion and Sociable Humanoid Robots," 123.

40. Breazeal, "Emotion and Sociable Humanoid Robots," 127.

41. Breazeal, "Emotion and Sociable Humanoid Robots," 125.

42. Breazeal, "Emotion and Sociable Humanoid Robots," 124.

43. Henri Bergson, *Creative Evolution*, translated by Arthur Mitchell (New York: Henry Holt, 1911), xii.

44. Bergson, *Creative Evolution*, xiii.

45. As Breazeal puts it, arousal (which ranges from high to low), valence (which ranges from positive to negative), and stance (which ranges from open to closed) triangulate what she terms the "affect space" in Kismet. She elaborates that "the current affective state occupies a single point in this space at a time. As the robot's affective state changes, this point moves about within this space. Note that this space not only maps to emotional states . . . but also to the level of arousal as well. . . . The affect space can be roughly partitioned into regions that map to each emotion process." Breazeal, "Emotion and the Sociable Humanoid Robots," 140.

46. Banu Subramaniam, *Ghost Stories for Darwin: The Science of Variation and the Politics of Diversity* (Urbana: University of Illinois Press, 2014).

47. Subramaniam, *Ghost Stories for Darwin*, 12.

48. Subramaniam, *Ghost Stories for Darwin*, 10.

49. Subramaniam, *Ghost Stories for Darwin*, 37.

50. Subramaniam, *Ghost Stories for Darwin*, 136.

51. Charles Darwin, *The Expression of the Emotions in Man and Animals* (New York: D. Appleton, 1897), 15.

52. Darwin, *The Expression of the Emotions*, 16–17.

53. Darwin, *The Expression of the Emotions*, 20–21.

54. Darwin, *The Expression of the Emotions*, 21.

55. Darwin, *The Expression of the Emotions*, 155.

56. Darwin, *The Expression of the Emotions*, 155.

57. Darwin, *The Expression of the Emotions*, 157.

58. Darwin, *The Expression of the Emotions*, 155.

59. Darwin, *The Expression of the Emotions*, 233.

60. Darwin, *The Expression of the Emotions*, 246.

61. Darwin, *The Expression of the Emotions*, 17.

62. Denise Ferreira da Silva, *Toward a Global Idea of Race* (Minneapolis: University of Minnesota Press, 2007), xv–xvi.

63. Da Silva, *Toward a Global Idea of Race*, 29.

64. Kyla Schuller, *The Biopolitics of Feeling: Race, Sex, and Science in the Nineteenth Century* (Durham NC: Duke University Press, 2018), 21.

65. Schuller, *The Biopolitics of Feeling*, 4.

66. See Cynthia Breazeal, "Social Interactions in HRI: The Robot View," *IEEE Transactions on Man, Cybernetics, and Systems* 20 (2003): 5, http://robotic.media.mit.edu/wp-content/uploads/sites/14/2015/01/Breazeal-IEEESMC-04-trv.pdf.

67. Cynthia Breazeal, Jesse Gray, and Matt Berlin, "An Embodied Cognition Approach to Mindreading Skills for Socially Intelligent Robots," *International Journal of Robotics Research* (2009): 656–80, 656.

68. Breazeal et al., "Social Interactions in HRI," 658, 663.

69. Sara Ahmed, *The Cultural Politics of Emotion* (New York: Routledge, 2004), 4.

70. Ahmed, *The Cultural Politics of Emotion*, 8–9.

71. Ahmed, *The Cultural Politics of Emotion*, 10.

72. Carla Hustak and Natasha Myers, "Involuntary Momentum: Affective Ecologies and the Sciences of Plant/Insect Encounters" *differences* 23 (2012): 74–118, 95.

73. Hustak and Myers, "Involuntary Momentum," 97.

74. Hustak and Myers, "Involuntary Momentum," 103

75. Errol Morris, "The Most Curious Thing," *New York Times*, May 19, 2008, https://opinionator.blogs.nytimes.com/2008/05/19/the-most-curious-thing/.

76. Morris, "The Most Curious Thing."

77. Infinite History Project MIT, "Cynthia Breazeal."

78. Elizabeth Wilson, *Affect and Artificial Intelligence* (Seattle: University of Washington Press, 2010), x.

79. Wilson, *Affect and Artificial Intelligence*, 115.

80. Wilson, *Affect and Artificial Intelligence*, 115.

81. Wilson, *Affect and Artificial Intelligence*, 55.

82. Wilson, *Affect and Artificial Intelligence*, 79.

83. Wilson, *Affect and Artificial Intelligence*, 81.

84. Wilson, *Affect and Artificial Intelligence*, 130.

85. Wilson, *Affect and Artificial Intelligence*, 130.

86. Katherine N. Hayles, *How We Became Posthuman: Virtual Bodies in Cybernetics, Literature, and Informatics* (Chicago: University of Chicago Press, 1999).

87. Maurice Merleau-Ponty, *The Phenomenology of Perception* (New York: Routledge, 2005), 75.

88. Merleau-Ponty, *The Phenomenology of Perception*, 141.

89. Donna Haraway, *When Species Meet* (Minneapolis: University of Minnesota Press, 2008), 4.

90. Haraway, *When Species Meet*, 4.

91. Haraway, *When Species Meet*, 62.

92. Ahmed, *The Cultural Politics of Emotion*, 10.

93. Kelly Dobson, "Machine Therapy," PhD diss., Massachusetts Institute of Technology, 2007, 22.

94. Dobson, "Machine Therapy," 98.

5. Machine Autonomy and the Unmanned Spacetime of Technoliberal Warfare

1. Kalindi Vora, "Limits of Labor: Accounting for Affect and the Biological in Transnational Surrogacy and Service Work," *South Atlantic Quarterly* 111 (2012): 681–700; Kalindi Vora, *Life Support: Biocapital and the New History of Outsourced Labor* (Minneapolis: University of Minnesota Press, 2015); Janet Jakobsen, "Perverse Justice," *GLQ* 18 (2012): 25.

2. Ian G. R. Shaw, "The Rise of the Predator Empire: Tracing the History of U.S. Drones," *Understanding Empire: Technology, Power, Politics*, accessed November 5, 2017, https://understandingempire.wordpress.com/2-o-a-brief-history-of-u-s-drones/.

3. Alice Ross, "Former US Drone Technicians Speak Out against Programme in Brussels," *The Guardian*, July 1, 2016, https://www.theguardian.com/world/2016/jul/01/us-drone-whistleblowers-brussels-european-parliament?CMP=share_btn_fb.

4. Lucy Suchman and Jutta Weber, "Human-Machine Autonomies Revised," February 13, 2015, https://www.researchgate.net/publication/272173538_Human-Machine_Autonomies_Revised, 8.

5. Suchman and Weber, "Human-Machine Autonomies Revised," 13.

6. Suchman and Weber, "Human-Machine Autonomies Revised," 14.

7. Lucy Suchman, *Human-Machine Reconfigurations: Plans and Situated Actions* (New York: Cambridge University Press, 2007), 141.

8. Suchman and Weber, "Human-Machine Autonomies Revised," 24.

9. Suchman and Weber, "Human-Machine Autonomies Revised," 2.

10. Adam Rothstein, *Drone* (New York: Bloomsbury, 2015), 82.

11. At the time, Boston Dynamics was a subsidiary of Google, but it began at MIT, where Marc Raibert and his team developed robots that could move and maneuver through space like animals (this was the Robotic Leg Lab).

12. Rothstein, *Drone*. See also Christopher Scholl, "US Army to Replace Human Soldiers with Humanoid Robots," *Global Research*, March 3, 2014, http://www.globalresearch.ca/us-army-to-replace-human-soldiers-with-humanoid-robots/5371657.

13. "The Mechanized Future of Warfare," *The Week*, December 13, 2014, http://theweek.com/articles/441634/mechanized-future-warfare.

14. Julian Smith, "Can Robots Be Programmed to Learn from Their Own Experiences?," *Scientific American*, March 23, 2009, http://www.scientificamerican.com/article/robot-learning/.

15. Boston Dynamics, "Atlas—The Agile Anthropomorphic Robot," accessed September 13, 2015, http://www.bostondynamics.com/robot_Atlas.html.

16. Angela Y. Davis, "The Approaching Obsolescence of Housework: A Working-Class Perspective," in *Women, Race, and Class*, by Angela Y. Davis (New York: Random House, 1981), 222–45.

17. Davis, "The Approaching Obsolescence of Housework."

18. Interview with Debbie Douglas, October 29, 2014 (emphasis ours).

19. Lisa Lowe, *The Intimacies of Four Continents* (Durham, NC: Duke University Press, 2015).

20. Denise Chow, "Boston Dynamics' New Atlas Robot Can't Be Pushed Around," *Yahoo Tech*, February 25, 2016, https://www.yahoo.com/tech/boston-dynamics-atlas -robot-cant-pushed-around-video-141752175.html.

21. Despina Kakoudaki, *Anatomy of a Robot: Literature, Cinema, and the Cultural Work of Artificial People* (New Brunswick, NJ: Rutgers University Press, 2014).

22. Anne McClintock, *Imperial Leather: Race, Gender, and Sexuality in the Colonial Contest* (New York: Routledge, 1995); Ann Laura Stoler, *Carnal Knowledge and Imperial Power: Race and the Intimate in Colonial Rule* (Berkeley: University of California Press, 2002).

23. Boston Dynamics, "Auralnauts Horror Edition," video, February 26, 2016, https:// www.youtube.com/watch?v=sAmyZP-qbTE.

24. Boston Dynamics, "Google's Atlas Robot," video, February 23, 2016, https://www .youtube.com/watch?v=9kawY1SMfYo.

25. Boston Dynamics, "Google's Atlas Robot."

26. Chow, "Boston Dynamics' New Atlas Robot."

27. Gregoire Chamayou, *Theory of the Drone*, translated by Janet Lloyd (New York: New Press, 2014), 23.

28. Neda Atanasoski, *Humanitarian Violence: The US Deployment of Diversity* (Minneapolis: University of Minnesota Press, 2013).

29. Chamayou, *Theory of the Drone*, 21.

30. Cited in Chamayou, *Theory of the Drone*, 21.

31. Chamayou, *Theory of the Drone*, 21.

32. Chamayou, *Theory of the Drone*, 22 (emphasis ours).

33. "'Drones' Will Dive into Atomic Blast; Blandy Reports Navy-AAF Plan to Use Robot Planes to Get Data in Pacific Test," *New York Times*, February 16, 1946, 9.

34. "NASA Picks Philco to Study Robot Laboratory for Mars," *New York Times*, July 23, 1964, 24.

35. Katherine Chandler, "American Kamikaze: Television-Guided Assault Drones in World War II," in *Life in the Age of Drone Warfare*, edited by Lisa Parks and Caren Kaplan (Durham, NC: Duke University Press, 2017), 90.

36. Chandler, "American Kamikaze," 97.

37. "SandFlea," accessed November 5, 2017, http://bostondynamics.com/robot _sandflea.html.

38. "Operation Sand Flea," last modified December 28, 2015, https://en.wikipedia .org/wiki /wiki/Operation_Sand_Flea.

39. Matt Simon, "Boston Dynamics' New Rolling, Leaping Robot Is an Evolutionary Marvel," *Wired*, March 1, 2017, https://www.wired.com/2017/03/boston-dynamics-new -rolling-leaping-robot-evolutionary-marvel/.

40. Simon, "Boston Dynamics."

41. Simon, "Boston Dynamics."

42. Rothstein, *Drone*.

43. Chamayou, *Theory of the Drone*, 38.

44. Chamayou, *Theory of the Drone*, 38–44.

45. Rothstein, *Drone*, 127.

46. Suchman, *Human-Machine Reconfigurations*.

47. Rothstein, *Drone*, 127–28.

48. Keith Feldman, "Empire's Verticality: The Af/Pak Frontier, Visual Culture, and Racialization from Above," *Comparative American Studies* 9 (2011): 325–41.

49. Feldman, "Empire's Verticality," 325.

50. Feldman, "Empire's Verticality," 331.

51. Neel Ahuja, *Bioinsecurities: Disease Interventions, Empire, and the Government of Species* (Durham, NC: Duke University Press, 2016), viii.

52. Katherine F. Chandler, "A Drone Manifesto," *Catalyst: Feminism, Theory, Technoscience* 2 (2016): 1–23, accessed November 5, 2017, http://catalystjournal.org/ojs/index.php/catalyst/article/view/chandler/165.

53. Jordan Crandall, "Materialities of the Robotic," in *Life in the Age of Drone Warfare*, edited by Lisa Parks and Caren Kaplan (Durham, NC: Duke University Press, 2017), 327–28.

54. Crandall, "Materialities of the Robotic," 328–29.

55. "Natalie Jeremijenko," interview by Lenny Simon, Center for the Study of the Drone at Bard College, November 13, 2013, http://dronecenter.bard.edu/interview-natalie-jeremijenko/.

56. "Natalie Jeremijenko."

57. "The Art of Drone Painting: A Conversation with Addie Wagenknecht," interview by Arthur Holland Michel, Center for the Study of the Drone at Bard College, December 6, 2014, http://dronecenter.bard.edu/art-of-drone-painting/.

58. Anjali Nath, "Touched from Below: On Drones, Screens, and Navigation," *Visual Anthropology* 29 (2016): 315–30, 322.

59. Peter Asaro, "The Labor of Surveillance and Bureaucratized Killing: New Subjectivities of Military Drone Operators," *Social Semiotics* (2013): 1–29, accessed November 5, 2017, doi: 10.1080/10350330.2013.777591.

60. Murtaza Hussein, "Former Drone Operators Say They Were Horrified by Cruelty of the Assassination Program," *The Intercept*, November 19, 2015, http://www.sott.net/article/306793-Former-drone-operators-say-they-were-horrified-by-cruelty-of-assassination-program.

61. Hussein, "Former Drone Operators Horrified."

62. Hussein, "Former Drone Operators Horrified," 2.

63. Hussein, "Former Drone Operators Horrified," 3.

64. Hussein, "Former Drone Operators Horrified," 3.

65. Ed Pilkington, "Life as a Drone Operator: 'Ever Step on Ants and Never Give It Another Thought?,'" *The Guardian*, November 19, 2015, http://www.theguardian.com/world/2015/nov/18/life-as-a-drone-pilot-creech-air-force-base-nevada.

66. Pilkington, "Life as a Drone Operator."

67. Heather Linebaugh, "I Worked on the US Drone Program. The Public Should Know What Really Goes On," *The Guardian*, December 29, 2013, http://www .theguardian.com/commentisfree/2013/dec/29/drones-us-military.

68. Linebaugh, "I Worked on the US Drone Program."

69. "'I Couldn't Stop Crying'—Ethan McCord Relives Collateral Murder Video," *Corrente*, April 7, 2013, http://www.correntewire.com/i_couldnt_stop_crying_ethan _mccord_relives_collateral_murder_video.

70. Personal correspondence with authors, October 2016.

6. Killer Robots

1. Campaign to Stop Killer Robots, "About Us," accessed November 9, 2017, www .stopkillerrobots.org/about-us/.

2. "Universal Declaration of Human Rights," United Nations General Assembly, December 10, 1948, http://www.un.org/en/universal-declaration-human-rights/; Jack Donnelly, *Universal Human Rights in Theory and Practice* (Ithaca, NY: Cornell University Press, 2013); Lynn Hunt, *Inventing Human Rights: A History* (New York: W. W. Norton, 2008).

3. Donnelly, *Universal Human Rights*, 15.

4. See for instance Samera Esmeir, *Juridical Humanity: A Colonial History* (Stanford, CA: Stanford University Press, 2012).

5. Katy Waldman, "Are Soldiers Too Emotionally Attached to Military Robots?," *Slate*, September 20, 2013, http://www.slate.com/blogs/futuretense/2013/09/20 /military_bots_inspire_strong_emotional_connections_in_troops_is_that_bad.html; Megan Garber, "Funerals for Fallen Robots," *The Atlantic*, September 20, 2013, https:// www.theatlantic.com/technology/archive/2013/09/funerals-for-fallen-robots/279861/.

6. Waldman, "Are Soldiers Too Emotionally Attached?"

7. Julie Carpenter, *Culture and Human Robot Interaction in Militarized Spaces: A War Story* (New York: Routledge, 2016).

8. Brent Rose, "The Sad Story of a Real Life R2-D2 Who Saved Countless Human Lives and Died," *Gizmodo*, December 28, 2011, accessed November 5, 2017, http:// gizmodo.com/5870529/the-sad-story-of-a-real-life-r2-d2-who-saved-countless-human -lives-and-died.

9. Saidiya Hartman, *Scenes of Subjection: Terror, Slavery, and Self-Making in Nineteenth-Century America* (New York: Oxford University Press, 1997).

10. Hartman, *Scenes of Subjection*, 18.

11. Hartman, *Scenes of Subjection*, 19.

12. Hartman, *Scenes of Subjection*, 20.

13. Hartman, *Scenes of Subjection*, 20.

14. David Goldman, "Google Wants You to Kick This Robot Puppy," CNN, February 10, 2015, accessed November 5, 2017, http://money.cnn.com/2015/02/10 /technology/google-robot-dog-spot/.

15. Boston Dynamics. "BigDog," accessed November 5, 2017, http://www .bostondynamics.com/robot_bigdog.html.

16. Goldman, "Google Wants You to Kick This Robot Puppy."

17. Goldman, "Google Wants You to Kick This Robot Puppy."

18. Goldman, "Google Wants You to Kick This Robot Puppy."

19. Samera Esmeir, *Juridical Humanity: A Colonial History* (Stanford, CA: Stanford University Press, 2012), 111.

20. Esmeir, *Juridical Humanity*, 112.

21. For a discussion of how care for animals has been legislated and utilized for a twentieth-century US imperial knowledge project, see Julietta Hua and Neel Ahuja, "Chimpanzee Sanctuary: 'Surplus Life' and the Politics of Transspecies Care," *American Quarterly* 65, no. 3 (2013): 619–37.

22. Sharon Sliwinski, *Human Rights in Camera* (Chicago: University of Chicago Press, 2011), 19.

23. Sliwinski, *Human Rights in Camera*, 21.

24. Sliwinski, *Human Rights in Camera*, 23.

25. Sliwinski, *Human Rights in Camera*, 25.

26. On the coloniality of the human rights regime, see Randall Williams, *The Divided World: Human Rights and Its Violence* (Minneapolis: University of Minnesota Press, 2010).

27. For more on the institution of humanitarian intervention as a tool of US empire building, see Neda Atanasoski, *Humanitarian Violence: The US Deployment of Diversity* (Minneapolis: University of Minnesota Press, 2013).

28. Lisa Lowe, *The Intimacies of Four Continents* (Durham, NC: Duke University Press, 2015), 6.

29. Campaign to Stop Killer Robots, "The Problem," accessed November 9, 2017, http://www.stopkillerrobots.org/the-problem/.

30. Campaign to Stop Killer Robots, "The Solution," accessed November 9, 2017, http://www.stopkillerrobots.org/the-solution/.

31. Human Rights Watch and International Human Rights Clinic, "Advancing the Debate on Killer Robots: 12 Key Arguments for a Preemptive Ban on Fully Autonomous Weapons," May 13, 2014, 21, https://www.hrw.org/news/2014/05/13/advancing-debate -killer-robots.

32. Human Rights Watch and International Human Rights Clinic, "Advancing the Debate on Killer Robots," 5.

33. Human Rights Watch and International Human Rights Clinic, "Advancing the Debate on Killer Robots," 10.

34. Human Rights Watch and International Human Rights Clinic, "Shaking the Foundations: The Human Rights Implications of Killer Robots," May 12, 2014, https:// www.hrw.org/report/2014/05/12/shaking-foundations/human-rights-implications-killer -robots (emphasis ours).

35. Nadia Prupis, "Legal Experts Raise Alarm over Shocking Use of 'Killer Robot' in Dallas," *Common Dreams*, July 8, 2016, https://www.commondreams.org/news/2016/07 /08/legal-experts-raise-alarm-over-shocking-use-killer-robot-dallas.

36. Prupis, "Legal Experts Raise Alarm."

37. Prupis, "Legal Experts Raise Alarm."

38. See also Marjorie Cohn, *Drones and Targeted Killing* (Northampton, MA: Olive Branch Press, 2015; 2017).

39. Hugh Gusterson, *Drones: Remote Control Warfare* (Cambridge, MA: MIT Press, 2016), 61.

40. Talal Asad, "Reflections on Violence, Law, and Humanitarianism," *Critical Inquiry* 41 (2015): 390–427, doi: 10.1086/679081.

41. Human Rights Watch, "Mind the Gap: The Lack of Accountability for Killer Robots," April 9, 2015, https://www.hrw.org/report/2015/04/09/mind-gap/lack-accountability-killer-robots.

42. Atanasoski, *Humanitarian Violence.*

43. Atanasoski, *Humanitarian Violence*, 170.

44. Human Rights Watch, "Losing Humanity: The Case against Killer Robots," May 12, 2014, 30, https://www.hrw.org/report/2014/05/12/shaking-foundations/human-rights-implications-killer-robots#page.

45. Atanasoski, *Humanitarian Violence*, 170.

46. Ronald Arkin, "Lethal Autonomous Systems and the Plight of the Non-Combatant," *AISB Quarterly* 137 (2013): 3.

47. Arkin, "Lethal Autonomous Systems," 1.

48. Gregoire Chamayou, *A Theory of the Drone*, translated by Janet Lloyd (New York: New Press, 2015), 208.

49. Chamayou, *A Theory of the Drone*, 208.

50. Chamayou, *A Theory of the Drone*, 208.

51. Chamayou, *A Theory of the Drone*, 208.

52. Chamayou, *A Theory of the Drone*, 208.

53. Vice, "The Dawn of Killer Robots," *Motherboard*, season 2, episode 25, https://motherboard.vice.com/en_us/article/bmjbqz/inhuman-kind-killer-robots.

54. Atanasoski, *Humanitarian Violence.*

55. Rick Cohen, "Humanitarian Aid Delivered by Drones: A New Frontier for NGOs?," *Nonprofit Quarterly*, July 16, 2014, https://nonprofitquarterly.org/2014/07/16/humanitarian-aid-delivered-by-drones-a-new-frontier-for-ngos/.

56. Kalev Leeataru, "How Drones Are Changing Humanitarian Disaster Response," *Forbes*, November 9, 2015, http://www.forbes.com/sites/kalevleetaru/2015/11/09/how-drones-are-changing-humanitarian-disaster-response/#132be0656cee.

57. Chamayou, *Theory of the Drone*, 136.

58. Chamayou, *Theory of the Drone*, 138.

59. Heather Linebaugh, "I Worked on the US Drone Program. The Public Should Know What Really Goes On," *The Guardian*, December 29, 2013, http://www.theguardian.com/commentisfree/2013/dec/29/drones-us-military.

60. Linebaugh, "I Worked on the US Drone Program."

61. Sophia Saifi, "Not a Bug Splat: Artists Give Drone Victims a Face in Pakistan," CNN, April 9, 2014, http://www.cnn.com/2014/04/09/world/asia/pakistan-drones-not-a-bug-splat/. Similar efforts to repersonify targets through visualizing them as human, or as in the latter example as human children, to overcome the erasure wrought by target status include a collective art project that focuses on connecting photographs of child

victims of drone strikes, Radical Art for These Times (RAFTT), "Child Drone Victims Memorial." The political project "Occupy Beale Airforce Base" obtained a list of young civilians killed in Pakistan and Yemen and creates portraits to document each name on the list, putting them together in one "quilt-like expression."

62. Asad, "Reflections on Violence."

63. Philip K. Dick, "The Defenders" (1961), Kindle ed. (Amazon Digital Services LLC, 2011), 10.

64. Dick, "The Defenders," 38–39.

65. Dick, "The Defenders," 39.

66. Dick, "The Defenders," 40.

67. Dick, "The Defenders," 42.

68. Chamayou, *Theory of the Drone*, 206.

69. IOCOSE, "In Times of Peace," 2014–16, https://docs.google.com/document/d/15-fK-oRpZZbPoH2JFjGUr8dYR3tLm30RFghUGUT-H9I/edit.

70. IOCOSE, "In Times of Peace."

71. Andrew Nunes, "Now Even Drones Are Taking Selfies," *Creators*, July 28, 2014, http://thecreatorsproject.vice.com/blog/now-even-drones-are-taking-selfies?tcpfbus.

72. Nunes, "Now Even Drones Are Taking Selfies."

Epilogue. On Technoliberal Desire

1. Zachary Canepari, Drea Cooper, and Emma Cott, "The Uncanny Lover," *New York Times*, video, June 11, 2015, https://www.nytimes.com/video/technology/100000003731634/the-uncannylover.html?action=click&contentCollection=technology&module=lede&region=caption&pgtype=article.

2. Emma Cott, "Sex Dolls That Talk Back," *New York Times*, Robotica, episode 5, June 11, 2015, https://www.nytimes.com/2015/06/12/technology/robotica-sex-robot-realdoll.html.

3. Canepari et al., "The Uncanny Lover."

4. Steven Huff, "Get Ready to Meet the First Sex Robots with Real Emotions," *Maxim*, January 29, 2017, https://www.maxim.com/maxim-man/first-sex-robot-real-emotions-2017-1.

5. Canepari et al., "The Uncanny Lover."

6. Lily Waddell, "Cyborgs with 'Warm Skin' for Humans to Practice on Hit Dating Scene," October 4, 2017, accessed December 1, 2017, https://www.dailystar.co.uk/news/latest-news/550418/Sex-robot-Tinder-practice-warm-skin-bedroom-bed-skills-human-relationships-advice-dating.

7. Waddell, "Cyborgs with 'Warm Skin.'"

8. Saidiya Hartman writes about this racial structure in the context of US slavery: "While the slave was recognized as a sentient being, the degree of sentience had to be cautiously calibrated in order to avoid intensifying the antagonisms of the social order." *Scenes of Subjection: Terror, Slavery, and Self-Making in Nineteenth-Century America* (New York: Oxford University Press, 1997), 93.

9. Alex Mar, "Love in the Time of Robots," *Wired*, October 17, 2017, https://www.wired.com/?p=2264647.

10. Mar, "Love in the Time of Robots."

11. Mar, "Love in the Time of Robots."

12. Mar, "Love in the Time of Robots."

13. John Berger, *Ways of Seeing* (London: British Broadcasting Corporation, 1972).

14. Zeynep Yenisey, "There's Now a Sex Robot with a G-Spot, So You Can Totally Give It an Orgasm," *Maxim*, March 22, 2017, accessed December 1, 2017, https://www.maxim.com/maxim-man/sex-robot-with-g-spot-2017-3.

15. Margi Murphy, "Built for Pleasure: Meet Samantha, the Artificially Intelligent Sex Robot Who 'REALLY Likes to Be Kissed,'" *The Sun*, March 17, 2017, accessed December 1, 2017, https://www.thesun.co.uk/tech/3115956/meet-samantha-and-artificially-intelligent-sex-robot-who-really-likes-to-be-kissed/.

16. Hortense J. Spillers, "Mama's Baby, Papa's Maybe: An American Grammar Book," *Diacritics* 17, no. 2 (1987): 67.

17. Spillers, "Mama's Baby, Papa's Maybe," 67.

18. Spillers, "Mama's Baby, Papa's Maybe," 67.

19. Spillers, "Mama's Baby, Papa's Maybe," 67.

20. Cheryl I. Harris, "Whiteness as Property," *Harvard Law Review* 106 (1994): 1709–91. See also Lisa Cacho, *Social Death: Racialized Rightlessness and the Criminalization of the Unprotected* (New York: New York University Press, 2012).

21. Spillers, "Mama's Baby, Papa's Maybe," 68.

22. Spillers, "Mama's Baby, Papa's Maybe," 76.

23. Elise Bohan, "This New Species of AI Wants to Be 'Superintelligent' When She Grows Up," *Big Think*, March 22, 2017, accessed June 5, 2017, http://bigthink.com/elise-bohan/the-most-human-ai-youve-never-heard-of-meet-luna.

24. The view of the human mind itself as a binary system is necessary for building intelligent computing systems, a belief which continues into the present. Elizabeth Wilson, *Affect and AI* (Seattle: University of Washington Press, 2010), 80.

25. Rodney Brooks, *Flesh and Machines: How Robots Will Change Us* (New York: Vintage, 2003), 35–36.

26. Luis Arana, email message to the authors, June 12, 2017.

27. Saidiya Hartman notes, "The law attempted to resolve the contradiction between the slave as property and the slave as person/laborer . . . by attending to the slave as both a form of property and a person. . . . The dual invocation of law (the black is a propertied person) designated the limits of rights of ownership and extended and constricted these rights as was necessary for the preservation of the institution [of slavery]." Hartman, *Scenes of Subjection*, 93–94.

Bibliography

Ahmed, Sara. *The Cultural Politics of Emotion.* New York: Routledge, 2004.

Ahuja, Neel. *Bioinsecurities: Disease Interventions, Empire, and the Government of Species.* Durham NC: Duke University Press, 2016.

Arendt, Hannah. *The Human Condition.* Chicago: University of Chicago Press, 1958.

Arkin, Ronald. "Lethal Autonomous Systems and the Plight of the Non-Combatant." *AISB Quarterly* 137 (2013): 1–9.

Asad, Talal. "Reflections on Violence, Law, and Humanitarianism." *Critical Inquiry* 41 (2015): 390–427. doi: 10.1086/679081.

Asaro, Peter. "The Labor of Surveillance and Bureaucratized Killing: New Subjectivities of Military Drone Operators." *Social Semiotics* (2013): 1–29. Accessed November 5, 2017. doi: 10.1080/10350330.2013.777591.

Atanasoski, Neda. *Humanitarian Violence: The U.S. Deployment of Diversity.* Minneapolis: University of Minnesota Press, 2013.

Atanasoski, Neda, and Kalindi Vora. "Postsocialist Politics and the Ends of Revolution." *Social Text* (2017): 1–16. doi: https://doi.org/10.1080/13504630.2017.1321712.

Atanasoski, Neda, and Kalindi Vora. "The Surrogate Effect: Technoliberalism and Whiteness in a 'Post' Labor Era." *Catalyst: Feminism, Theory, Technoscience* 4, no. 1 (2018). https://catalystjournal.org/index.php/catalyst/article/view/29637/pdf.

Atanasoski, Neda, and Kalindi Vora. "Surrogate Humanity: Posthuman Networks and the (Racialized) Obsolescence of Labor." *Catalyst: Feminism, Theory, Technoscience* 1, no. 1 (2015). http://catalystjournal.org/ojs/index.php/catalyst/article/view/ata_vora.

Barad, Karen. "Intra-Actions: Interview of Karen Barad by Adam Kleinmann." *Mousse* (2012) 34: 76–81.

Barad, Karen. *Meeting the Universe Halfway: Quantum Physics and the Entanglement of Matter and Meaning.* Durham, NC: Duke University Press, 2007.

Benjamin, Ruha. "Innovating Inequity: If Race Is a Technology, Postracialism Is the Genius Bar." *Ethnic and Racial Studies* 39 (2016). doi: 10.1080/01419870.2016.1202423.

Bergson, Henri. *Creative Evolution*. Translated by Arthur Mitchell. New York: Henry Holt, 1911.

Breazeal, Cynthia. "Emotion and Sociable Humanoid Robots." *International Journal of Human–Computer Studies* 59 (2003): 119–55.

Breazeal, Cynthia. "Social Interactions in HRI: The Robot View." *IEEE Transactions on Man, Cybernetics, and Systems* 20 (2003): 1–6. http://robotic.media.mit.edu/wp-content/uploads/sites/14/2015/01/Breazeal-IEEESMC-04-trv.pdf.

Breazeal, Cynthia, Jesse Gray, and Matt Berlin. "An Embodied Cognition Approach to Mindreading Skills for Socially Intelligent Robots." *International Journal of Robotics Research* (2009): 656–80.

Brooks, Rodney. *Flesh and Machines: How Robots Will Change Us*. New York: Vintage, 2003.

Bruce, Allison, Illah Nourbakhsh, and Reid Simmons. "The Role of Expressiveness and Attention in Human-Robot Interaction." *Proceedings of the IEEE International Conference on Robotics and Automation*, Washington, DC, May 2002. http://www.cs.cmu.edu/~abruce/papers/bruce_icra2002.pdf.

Buck-Morss, Susan. *Dreamworld and Catastrophe: The Passing of Mass Utopias East and West*. Cambridge, MA: MIT Press, 2002.

Cacho, Lisa Marie. *Social Death: Racialized Rightlessness and the Criminalization of the Unprotected*. New York: New York University Press, 2012.

Carpenter, Julie. *Culture and Human Robot Interaction in Militarized Spaces: A War Story*. New York: Routledge, 2016.

Chakrabarty, Dipesh. *Provincializing Europe: Postcolonial Thought and Historical Difference*. Princeton, NJ: Princeton University Press, 2000.

Chamayou, Gregoire. *Theory of the Drone*. Translated by Janet Lloyd. New York: New Press, 2014.

Chandler, Katherine. "American Kamikaze: Television-Guided Assault Drones in World War II." In *Life in the Age of Drone Warfare*, edited by Lisa Parks and Caren Kaplan, 89–111. Durham, NC: Duke University Press, 2017.

Chandler, Katherine. "A Drone Manifesto." *Catalyst: Feminism, Theory, Technoscience* 2 (2016): 1–23. Accessed November 5, 2017. http://catalystjournal.org/ojs/index.php/catalyst/article/view/chandler/165.

Chun, Wendy Hui Kyong. "Introduction: Race and/as Technology, or How to Do Things to Race." *Camera Obscura* 24, no. 1 (2009): 7–35. doi: 10.1215/02705346–2008–013.

Chun, Wendy Hui Kyong. "On Software, or the Persistence of Visual Knowledge." *grey room* 18 (2005): 26–51.

Chun, Wendy Hui Kyong. "Race and/as Technology, or How to Do Things to Race." In *Race after the Internet*, edited by Lisa Nakamura and Peter A. Chow-White, 38–60. New York: Routledge, 2011.

Cohn, Marjorie. *Drones and Targeted Killing*. Northampton, MA: Olive Branch Press, [2015] 2017.

Coleman, Beth. "Race as Technology." *Camera Obscura* 24, no. 1 (2009): 177–207. doi: 10.1215/02705346–2008–018.

Crandall, Jordan. "Materialities of the Robotic." In *Life in the Age of Drone Warfare*, edited by Lisa Parks and Caren Kaplan, 324–43. Durham, NC: Duke University Press, 2017.

Darling, Kate. "Extending Legal Protection to Social Robots: The Effects of Anthropomorphism, Empathy, and Violent Behavior towards Robotic Objects." In *Robot Law*, edited by Ryan Calo, A. Michael Froomkin, Laurie Silvers, Mitchell Rubenstein, and Ian Kerr. Cheltenham, UK: Edward Elgar, 2016. http://papers.ssrn.com/s013/papers.cfm?abstract_id=2044797.

Darling, Kate. "Who's Johnny? Anthropomorphic Framing in Human-Robot Interaction, Integration, and Policy." In *Robot Ethics 2.0*, edited by P. Lin, G. Bekey, K. Abney, and R. Jenkins. Oxford: Oxford University Press, forthcoming.

Darwin, Charles. *The Expression of the Emotions in Man and Animals*. New York: D. Appleton, 1897.

da Silva, Denise Ferreira. *Toward a Global Idea of Race*. Minneapolis: University of Minnesota Press, 2007.

Dautenhahn, Kerstin. "Design Issues on Interactive Environments for Children with Autism." *Proceedings of the Third International Conference on Disability, Virtual Reality and Associated Technologies*, University of Reading (2000): 153–61. web.mit.edu/16.459/www/Dautenhahn.pdf.

Dautenhahn, Kerstin. "Robots as Social Actors: AURORA and the Case of Autism." *Proceedings of the Third Cognitive Technology Conference* (1999): 1–16. http://citeseerx.ist.psu.edu/viewdoc/summary?doi=10.1.1.190.1767.

Davis, Angela. *Women, Race, and Class*. New York: Vintage, 1983.

Donnelly, Jack. *Universal Human Rights in Theory and Practice*. Ithaca, NY: Cornell University Press, 2013.

Ellison, Ralph. "The Negro and the Second World War." In *Cultural Contexts for Ralph Ellison's Invisible Man*, edited by Eric J. Sindquist. New York: St. Martin's, 1995.

El-Tayeb, Fatima. "Black Women in Antiquity, edited by Ivan Van Sertima, 1988." *Contemporaryand.com* (forthcoming).

Esmeir, Samera. *Juridical Humanity: A Colonial History*. Stanford, CA: Stanford University Press, 2012.

Fanon, Frantz. *The Wretched of the Earth*. New York: New Grove, 1967.

Federici, Silvia. *Caliban and the Witch: Women, the Body, and Primitive Accumulation*. New York: Automedia, 2004.

Feldman, Keith P. "Empire's Verticality: The Af/Pak Frontier, Visual Culture, and Racialization from Above." *Comparative American Studies: An International Journal* 9, no. 4 (2011): 325–41.

Fortunati, Leopoldina. *The Arcane of Reproduction: Housework, Prostitution, Labor and Capital*. Translated by Hilary Creek. Edited by Jim Fleming. New York: Autonomedia, 1989.

Gibson-Graham, J. K. *A Postcapitalist Politics*. Minneapolis: University of Minnesota Press, 2006.

Goldberg, David Theo. *Racist Culture: Philosophy and the Politics of Meaning*. Oxford: Blackwell, 1993.

Gordon, Avery. "Preface." In *An Anthropology of Marxism*, by Cedric Robinson. Hampshire, UK: Ashgate, 2001.

Gusterson, Hugh. *Drones: Remote Control Warfare*. Cambridge, MA: MIT Press, 2016.

Haraway, Donna. *A Cyborg Manifesto: Science, Technology and Socialist Feminism in the Late Twentieth Century*. New York: Routledge, 1991.

Haraway, Donna. *ModestWitness@Second_Millennium: FemaleMan_Meets_OncoMouse: Feminism and Technoscience*. New York: Routledge, 1997.

Haraway, Donna. *When Species Meet*. Minneapolis: University of Minnesota Press, 2007.

Harris, Cheryl I. "Whiteness as Property." *Harvard Law Review* 106 (1994): 1707–91.

Hartman, Saidiya. *Scenes of Subjection: Terror, Slavery, and Self-Making in Nineteenth Century America*. New York: Oxford University Press, 1997.

Hayles, Katherine N. *How We Became Posthuman: Virtual Bodies in Cybernetics, Literature, and Informatics*. Chicago: University of Chicago Press, 1999.

Herzig, Rebecca, and Banu Subramaniam. "Labor in the Age of 'Bio-Everything.'" *Radical History Review*, no. 127 (2017): 103–24.

Hong, Grace Kyungwon. "Existentially Surplus Women of Color Feminism and the New Crises of Capitalism." *GLQ: A Journal of Lesbian and Gay Studies* 18, no. 1 (2012): 87–106.

Hua, Julietta, and Neel Ahuja. "Chimpanzee Sanctuary: 'Surplus' Life and the Politics of Transspecies Care." *American Quarterly* 65, no. 3 (2013): 619–37.

Hua, Julietta, and Kasturi Ray. "Beyond the Precariat: Race, Gender, and Labor in the Taxi and Uber Economy." *Social Identities: Journal for the Study of Race, Nation, and Culture* 24, no. 2 (2018): 271–89.

Hunt, Lynn. *Inventing Human Rights: A History*. New York: W. W. Norton, 2008.

Hustak, Carla, and Natasha Myers. "Involutionary Momentum: Affective Ecologies and the Sciences of Plant/Insect Encounters." *differences* 23, no. 3 (2012): 74–118.

Irani, Lilly. "The Cultural Work of Microwork." *New Media and Society* 17, no. 5 (2013): 720–39.

Jakobsen, Janet. "Perverse Justice." *GLQ: A Journal of Lesbian and Gay Studies* 18, no. 1 (2012): 19–45.

Kakoudaki, Despina. *Anatomy of a Robot: Literature, Cinema, and the Cultural Work of Artificial People*. New Brunswick, NJ: Rutgers University Press, 2014.

Lewis, Sophie. "Cthulu Plays No Role for Me." *Viewpoint*, May 8, 2017.

Licklider, Joseph C. R. "Man–Computer Symbiosis." *IRE Transactions on Human Factors in Electronics* 1 (1960): 4–11.

Light, Jennifer S. "When Women Were Computers." *Technology and Culture* 40, no. 3 (July 1999): 455–83.

Lowe, Lisa. "History Hesitant," *Social Text* 33, no. 4 (2015): 85–107.

Lowe, Lisa. *The Intimacies of Four Continents*. Durham, NC: Duke University Press, 2015.

Marez, Curtis. *Farm Worker Futurism: Speculative Technologies of Resistance*. Minneapolis: University of Minnesota Press, 2016.

Martín-Cabrera, Luis. "The Potentiality of the Commons: A Materialist Critique of Cognitive Capitalism from the Cyberbracer@s to the Ley Sinde." *Hispanic Review* 80, no. 4 (2012): 583–605.

Marx, Karl. "The Fragment on Machines." *Grundrisse*. Accessed November 17, 2017. thenewobjectivity.com/pdf/marx.pdf.

Marx, Karl, and Friedrich Engels. "Community Manifesto: Manifesto of the Communist Party." In *Marx/Engels Selected Works*, vol. 1. Moscow: Progress, 1969.

McClintock, Anne. *Imperial Leather: Race, Gender, and Sexuality in the Colonial Contest*. New York: Routledge, 1995.

Melamed, Jodi. "Racial Capitalism," *Critical Ethnic Studies* 1. no. 1 (2015): 76–85.

Melamed, Jodi. *Represent and Destroy: Rationalizing Violence in the New Racial Capitalism*. Minneapolis: University of Minnesota Press, 2011.

Melamed, Jodi. "The Spirit of Neoliberalism: From Racial Liberalism to Neoliberal Multiculturalism." *Social Text* 24, no. 4 (2006): 1–24.

Memmi, Albert. *The Colonizer and the Colonized*. London: Orion, 1965.

Merleau-Ponty, Maurice. *The Phenomenology of Perception*. New York: Routledge, 2005.

Mies, Maria. *Patriarchy and Accumulation on a World Scale: Women in the International Division of Labour*. London: Zed, 1986.

Minsky, Marvin. "Steps toward Artificial Intelligence." *Proceedings of the IRE* 49 (1961): 8–30.

Mori, Masahiro. "The Uncanny Valley." *Energy* 7 (1970): 33–35. http://comp.dit.ie /dgordon/Courses/CaseStudies/CaseStudy3d.pdf.

Murphy, Michelle. *Seizing the Means of Reproduction: Entanglements of Feminism, Health, and Technoscience*. Durham, NC: Duke University Press, 2012.

Nakamura, Lisa. "Indigenous Circuits: Navajo Women and the Racialization of Early Electronic Manufacture." *American Quarterly* 66 (2014): 919–41.

Nath, Anjali. "Touched from Below: On Drones, Screens, and Navigation." *Visual Anthropology* 29 (2016): 315–30.

Penley, Constance, and Andrew Ross. "Cyborgs at Large: Interview with Donna Haraway." In *Technoculture*, edited by Constance Penley and Andrew Ross, 1–20. Minneapolis: University of Minnesota Press, 1991.

Rhee, Margaret. "In Search of My Robot: Race, Technology, and the Asian American Body." *Scholar and Feminist Online* 13.3–14.1 (2016). http://sfonline.barnard.edu /traversing-technologies/margaret-rhee-in-search-of-my-robot-race-technology-and -the-asian-american-body/.

Richardson, Kathleen. *An Anthropology of Robotics and AI: Annihilation Anxiety and Machines*. New York: Routledge, 2015.

Rifkin, Jeremy. *The Zero Marginal Cost Society: The Internet of Things, the Collaborative Commons, and the Eclipse of Capitalism*. New York: St. Martin's Griffin, 2014.

Robinson, Cedric. *Black Marxism: The Making of the Black Radical Tradition*. Chapel Hill: University of North Carolina Press, 2000.

Roh, David S., Betsy Huang, and Greta A. Niu. "Technologizing Orientalism: An Introduction." In *Techno-Orientalism: Imagining Asia in Speculative Fiction, History, and*

Media, edited by David S. Roh et al., 1–22. New Brunswick, NJ: Rutgers University Press, 2015.

Rose, David. *Enchanted Objects: Innovation, Design, and the Future of Technology.* New York: Scribner, 2015.

Rothstein, Adam. *Drone.* New York: Bloomsbury, 2015.

Schuller, Kyla. *The Biopolitics of Feeling: Race, Sex, and Science in the Nineteenth Century.* Durham, NC: Duke University Press, 2018.

Scott, David. *Conscripts of Modernity: The Tragedy of Colonial Enlightenment.* Durham, NC: Duke University Press, 2004.

Selisker, Scott. *Human Programming: Brainwashing, Automatons, and American Unfreedom.* Minneapolis: University of Minnesota Press, 2016.

Shaw, Ian G. R. "The Rise of the Predator Empire: Tracing the History of U.S. Drones." *Understanding Empire: Technology, Power, Politics.* Accessed November 5, 2017. https://understandingempire.wordpress.com/2-0-a-brief-history-of-u-s-drones.

Singh, Nikhil Pal. "Cold War." *Social Text* 27, no. 3 (2009): 67–70.

Singh, Nikhil Pal. *Race and America's Long War.* Berkeley: University of California Press, 2017.

Sliwinski, Sharon. *Human Rights in Camera.* Chicago: University of Chicago Press, 2011.

Spillers, Hortense. "Mama's Baby, Papa's Maybe: An American Grammar Book." *Diacritics* 17, no. 2 (1987): 64–81.

Spivak, Gayatri Chakravorty. *A Critique of Postcolonial Reason: Toward a History of the Vanishing Present.* Cambridge, MA: Harvard University Press, 1999.

Spivak, Gayatri Chakravorty. "Scattered Speculations on the Question of Value." In *In Other Worlds: Essays in Cultural Politics*, edited by Gayatri Chakravorty Spivak, 212–42. New York: Routledge, 1988.

Stern, Andy, with Lee Kravitz. *Raising the Floor: How a Universal Basic Income Can Renew Our Economy and Rebuild the American Dream.* New York: Public Affairs, 2016.

Stoler, Ann Laura. *Carnal Knowledge and Imperial Power: Race and the Intimate in Colonial Rule.* Berkeley: University of California Press, 2002.

Stoler, Ann Laura. *Race and the Colonial Education of Desire: Foucault's History of Sexuality and the Colonial Order of Things.* Durham, NC: Duke University Press, 1995.

Stoler, Ann Laura. "Tense and Tender Ties: The Politics of Comparison in North American History and (Post) Colonial Studies." In *Haunted by Empire: Geographies of Intimacy in North American History*, edited by Ann Laura Stoler, 23–70. Durham, NC: Duke University Press, 2006.

Subramaniam, Banu. *Ghost Stories for Darwin: The Science of Variation and the Politics of Diversity.* Urbana: University of Illinois Press, 2014.

Suchman, Lucy. "Figuring Personhood in Sciences of the Artificial." Lancaster, UK: Department of Sociology, Lancaster University, 2004. http://www.comp.lancs.ac.uk/sociology/papers/suchman-figuring-personhood.pdf.

Suchman, Lucy. *Human-Machine Reconfigurations: Plans and Situated Actions.* Cambridge: Cambridge University Press, 2006.

Suchman, Lucy, and Jutta Weber. "Human-Machine Autonomies Revised." Symposium on Autonomous Weapons Systems—Law, Ethics, Policy, European University

Institute, Florence, April 24–25, 2015. https://www.researchgate.net/publication
/272173538_Human-Machine_Autonomies_Revised.

Terranova, Tiziana. *Network Culture: Politics for the Information Age*. London: Pluto, 2004.

United Nations General Assembly. "Universal Declaration of Human Rights." December 10, 1948. http://www.un.org/en/universal-declaration-human-rights/.

Vora, Kalindi. *Life Support: Biocapital and the New History of Outsourced Labor*. Minneapolis: University of Minnesota Press, 2015.

Vora, Kalindi. "Limits of 'Labor': Accounting for Affect and the Biological in Transnational Surrogacy and Service Work." *South Atlantic Quarterly* 111, no. 4 (2012): 681–700.

Vora, Kalindi. "The Transmission of Care: Affective Economies and Indian Call Centers." In *Intimate Labors: Cultures, Technologies, and the Politics of Care*, edited by Eileen Boris and Rhacel Salazar Parreñas, 33–48. Stanford, CA: Stanford University Press, 2010.

Weheliye, Alexander. *Habeas Viscus: Racializing Assemblages, Biopolitics, and Black Feminist Theories of the Human*. Durham, NC: Duke University Press, 2014.

Williams, Randall. *The Divided World: Human Rights and Its Violence*. Minneapolis: University of Minnesota Press, 2010.

Wilson, Elizabeth. *Affect and Artificial Intelligence*. Seattle: University of Washington Press, 2010.

Winner, Langdon. "Do Artifacts Have a Politics?" *Daedalus* 109 (1980): 121–36.

Wynter, Sylvia. "Unsettling the Coloniality of Being/Power/Truth/Freedom: Towards the Human, after Man, Its Overrepresentation—An Argument." CR: *The New Centennial Review* 3, no. 3 (2003): 257–337.

Wynter, Sylvia, and Katherine McKittrick. "Unparalleled Catastrophe for Our Species? or, To Give Humanness a Different Future: Conversations." In *Sylvia Wynter: On Being Human as Praxis*, edited by Katherine McKittrick, 9–89. Durham, NC: Duke University Press, 2015.

Index

Page numbers followed by *f* indicate illustrations.

surrogate human effect, 5; of AMT, 99; of empathy, 166–72; in field of war, 136–39, 149–51, 156, 157; of technoliberalism, 5–6, 8–12, 13, 17–19, 21–25, 28, 30–31, 58, 60, 70–71, 82, 90, 110–11, 136–38, 140, 153, 164–65

surrogate humanity, 6, 57, 90

surveillance, 150, 154, 160, 179–80, 185

TaskRabbit, 63, 92, 94

TaskUs, 102

Taylorism, 160

TechBoom 2.0, 2–4

technoliberal capitalism, 23, 59–64, 126, 128

technoliberalism: desire in, 6, 22, 189, 192, 194, 196; freedom as goal of, 57–58; racial capitalism and, 4, 6, 15, 24, 50, 51; racial engineering in, 5; racial grammar of, 5, 6, 8, 17, 22; repurposing tools of, 158, 159; socialist imaginary in, 24, 58, 63, 65, 80, 85; surrogate human effect of, 5–6, 8–12, 13, 17–19, 21–25, 28, 30–31, 58, 60, 70–71, 82, 90, 110–11, 136–38, 140, 153, 164–65; use of term, 4, 12

technological unemployment, use of term, 30

techno-Orientalism, 45–46

telechiric machines, 149–50

Terranova, Tiziana, 92

Third Industrial Revolution, 59, 65, 197n2

3D printing, 58, 68, 83–85

Tompkins, Silvan, 129

totalitarianism, mechanization of, 40

Trailers, Drake, 41

transparency, 92–93, 110, 120, 124–25, 128, 131, 132

Trump, Donald, 2; anti-immigrant policies, 49; on border wall, 46–47; as giant robot, 51f, 51–53; promises of jobs, 28; racist rhetoric of, 47–50, 53; white job loss and, 24, 47–49; white supremacy and, 13

Turing, Alan, 128

Turing test, 102

turkers. See Amazon Mechanical Turk

Turkopticon project, 106–7

Twilight Zone, The (television series), 38–39, 39f

UAV. See unmanned aerial vehicle

Uber, 63, 92, 102

UBI. See universal basic income

uncanny valley, 111, 188

unfreedom, 5, 8, 11–12; automation and, 40, 143; autonomy and, 10; desire for, 192, 194; dynamic with freedom, 11, 13–14, 23, 31, 33, 41–42, 143; of fascism, 40; illiberal states of, 11–12; racial capitalism and, 21, 30–31, 37, 38, 91, 191; slavery as, 32

Unimate (robot), 42f

universal basic income (UBI), 62–64

unmanned, myth of, 136–38, 156, 157–58

unmanned aerial vehicle (UAV), 137. See also drone warfare

Varde, Moshe, 60

vita activa, 31

Wagenknecht, Addie, 158

Walmart, 97–98

Walton, Sam, 97

war and warfare: machine autonomy in, 135–36, 139–42; robotic research for, 141–42; surrogate human effect in, 136–38, 149–51, 156, 157; technoliberal, 148–49, 155, 157–58, 168, 179. See also drone warfare

Weber, Jutta, 139

Weheliye, Alexander, 15

Westmoreland, Cian, 137

When Species Meet (Haraway), 131–32

white supremacy, 13, 14, 29, 46–47, 48, 50, 198n4

WikiLeaks, 162

Wilson, Elizabeth, 128–29

wind power, 134–35

Winner, Langdon, 56, 71

Wired (magazine), 63, 96, 152, 190, 207n47

witch hunts, 67–68

World Economic Forum, 27–28, 59, 197n2

World Social Forum, 85–86

World Trade Organization, 49

Wynter, Sylvia, 16, 136

Zapatistas, 85–86

Zero Marginal Cost Society, The (Rifkin), 65–66

Zuckerberg, Mark, 64

Printed in the USA
CPSIA information can be obtained
at www.ICGtesting.com
BVHW052114110823
668476BV00006B/18